# In Full View

Three Ways
of Seeing
California
Plants

NOT ALTOGETHER ABOUT PLANTS

Heyday Books    Berkeley, California

Co-published by Heyday Books and the Headlands Center for the Arts. Please address orders, inquiries, or correspondence to Heyday Books, Box 9145, Berkeley, California 94709.

Publisher's Cataloging in Publication
*(Prepared by Quality Books Inc.)*

Keator, Glenn.
   In full view ; three ways of seeing California plants  / Glenn Keator, Linda Yamane, Ann Lewis.
   p. cm.
   ISBN 0-930588-77-0

   1. Plants--California.  2. Plants--California--Pictorial works.
I. Yamane, Linda.  II. Lewis, Ann P., ill.  III. Title.

QK149.K43 1995                581.9'794
                              QB195-1458

Cover and interior design by Diane Burk
Printed in the United States of America
by Publisher's Press, Salt Lake City, Utah

10,  9,  8,  7,  6,  5,  4,  3,  2,  1

Headlands Center for the Arts, a non-profit corporation, works in partnership with the Golden Gate National Recreation Area. We are indebted to the James Irvine Foundation for its generous support of this project. *In Full View* also received support from The Leavens Foundation and the Fireman's Fund Company Foundation.

*Herbs. Leaves* simple to decompound, *alternate*; stipules none
with ... only on the ... ; leaves finely dissected or more than
or minute. *Flowers* small, *usually in umbels*, rarely in ...
headlike clusters, umbels simple or compound. *Calyx ...*
... *5-lobed; lobes inconspicuous. Petals* 5, on the ...
... 5, on the epigynous disk; *anthers versatile. Ovary ...*
styles 2, persistent, often on a conic or depressed style ...
... dry; *carpels* 2, *1-seeded*, with 9 or 5 chief ...
... 4 other smaller ribs, usually separating at ...
... of union (commissure); after ... borne
... axis (carpophore); ribs often ... ... usually
... ... (the first President.)

... ... depends upon ... oil tubes in the fruit for
... genera. To see these, cut a thin cross section
... knife and examine with the low power of
... The oil tubes are hollows just outside
... The key is given mostly to the genera only.
... doubtful whether beginners should go beyond the family.
(F. & ... pp. 271–290.)

A. **Leaves simple.**
B. Leaves awl-shaped to lanceolate or oblanceolate or oblong.
C. Leaves entire; flowers white or yellow, in umbels. GROUP ... (p. 165)
CC. Leaves lobed to dentate; flowers white or blue, ... somewhat spiny
heads. GROUP 1, A (p. 163)
BB. **Leaves ovate to ... or kidney-shaped.**
D. Marsh or water plants; leaves kidney-shaped, wider than long; umbel simple.
GROUP 3, B (p. 165)
D'D. Not marsh nor water plants; leaves ... or longer, at least longer than
wide; umbel compound. GROUP 1, B (p. 163)
... Leaves ... or very deeply dissected.
... or scaly. GROUP ...
... none or almost none ... GROUP 2 (p. 164)
... dorsally; lateral ribs more or less ... GROUP 3 (p. 165)
... dorsally, usually somewhat flattened laterally. GROUP 4 (p. 166)
... interval.
... Leaves many times compound. GROUP 1
... prickly heads ... (Said to be from Gk. ...) 
... a remedy for flatulency.) *Eryngium* ...
... in compound umbels, the umbellets often in headlike clusters ...
... or hardly bracted.

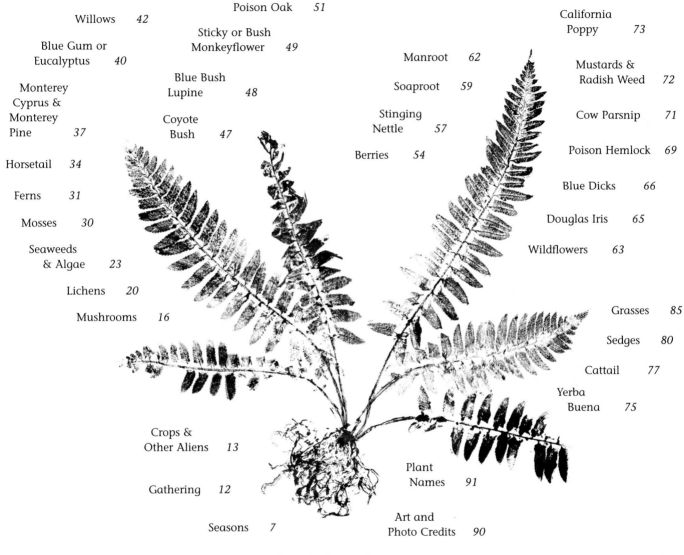

# Introduction

❦

Sometimes the wind stirs the leaves on trees and they rustle audibly. Branches might creak, and once in a while you can even hear the popping of seeds as they explode from their pods. But these are exceptional incidents. Mostly, it seems, plants are silent—wonderfully, fully silent. Perhaps, one speculates, they are secure in self-knowledge.

Or maybe they announce themselves so dramatically by their intense smells, colors, textures, and tastes that sound is unnecessary.

This is hardly the case with humans. We are full of squawks and squeals, hums and hisses, grunts and groans, laughter and cheers. Our pain, our joy, our fear, our awe, our passion, our curiosity, even our idleness all seem incomplete without the accompaniment of appropriate sounds.

For the last two years a small group of this noisy species has been getting together to create a book about

California plants. The group consisted of Glenn Keator, a California-based botanist and nature writer; Ann Lewis, a visual artist whose work often incorporates plant materials; Linda Yamane, a writer, artist, basketweaver, and tribal scholar of Rumsien Ohlone descent; Diane Burk, a graphic designer; and Malcolm Margolin, a writer and publisher of Heyday Books.

The team was assembled, supported, coaxed, and inspired by staff at Headlands Center for the Arts, which sponsored the project. Jennifer Dowley, former executive director, and Ann Chamberlain, former program director, oversaw the first phases of the project; it was completed under the direction of Kathryn Reasoner, executive director, Fritzie Brown, director of operations, and Donna Graves, program director. Along the way, crucial assistance was provided by anthropologist Vicki Patterson, Kashaya Pomo horticulturist Vana Lawson, Bodega Miwok scholar David Peri, and Golden Gate National Recreation Area Park Curator Diane Nicholson.

The Headlands Center for the Arts is located at the Marin Headlands just north of San Francisco and is part of the Golden Gate National Recreation Area of the National Park Service. One of the Center's goals is to serve as a laboratory for artists to investigate our connections to place, using the Marin Headlands as a site for research and exploration. Visual, literary, performance, and media artists from around the country and abroad live and maintain studio space in the historic, former military structures of the

Headlands. The combination of a spectacular location and supportive facilities has served since 1982 as a foundation for probing new ways of looking at the creative process, connecting with diverse audiences, and exploring the natural and historic environment of the Marin Headlands.

With miles of seashore and lagoon, meadows and brushland, freshwater marshes and willow-lined creek, and its groves of exotic trees, the Marin Headlands offered our small group much variety, challenge, and delight. Lying more or less in the middle of California's coastal region, the Headlands hosts plants generally found farther to the

north and south, and plants not native to California at all that have become well-established and widespread.

At meeting after meeting, held strategically around noon so that we could get the Center's chef, Laurie MacKenzie, to make us lunch, we talked and thought, drew up lists of plants, tore up lists of plants, misplaced lists of plants, and made new lists of plants. We struggled with how to organize these lists: whether, for example, to create separate sections for strawberries, blackberries, etc., or lump them all together under berries. We wrestled with the design and title of the book and how to balance the different contributions. We scheduled meetings,

postponed meetings, held meetings, and in the end became good friends. Almost, it seems, as a byproduct of that friendship, we also may have produced a book of great beauty and importance.

*In Full View: Three Ways of Seeing California Plants* presents three very different perspectives—each given full play, each amplifying rather than contradicting each other. Glenn brought to the book his considerable knowledge not only of botany, but of history as well. Linda drew upon her own experiences with plants and the ways in which those experiences have been shaped by thousands of years of knowledge, and of economic and emotional ties that native people have developed with the plants of California. Ann brought to the plants and to the book the eye of an artist and an open willingness to experiment with new media that resulted in a surprisingly fresh vision. Diane wove these varied perspectives into an inviting and beautiful whole. Malcolm and Donna offered their own perspectives, which included a bit of hand-wringing and brow-furrowing over practical concerns, and a sense that it was a great privilege to be working with some of the finest people we had ever met. We all shared the feeling, as the book unfolded and developed, that the plants of California are among the most fascinating and compelling of life forms, worthy of being seen from many perspectives, from a fuller (if never ultimately full) view.

🌾

*Malcolm Margolin*
Heyday Books

*Donna Graves*
Headlands Center
for the Arts

# Seasons

*Glenn Keator*  🌿

The typical temperate climate, occurring across most of the United States and Europe, experiences at least four distinct seasons. But in California, everyone knows, this concept is less aptly applied—at least in the mild climates of low elevations.

Lowland California has what is called a Mediterranean climate: mild, rainy winters; intermittently wet springs; and hot, dry, rainless summers and early falls. Only the Mediterranean Basin, the Cape Province of South Africa, central Chile, and southwestern Australia have comparable climates. And our coastal climates are modified even more, for the proximity of cold water creates fog in summer so that even though summers are rainless, they are also cool and humid. Where coastal forests occur, fog condenses on branches to drip to the ground as extra precipitation! So coastal California has a fairly nonseasonal climate—or so the figures on temperature and precipitation would suggest—with cool conditions year round.

Nonetheless coastal California has its own special rhythms.  Let's begin with the start of the rainy season from mid-October or late November to early December. Short day length signals a change from warm to cool daytime temperatures. As days cool, nights turn even cooler, sometimes cold enough so that by December and January we experience several nights of freezing.

During this period plants take advantage of the rains, yet the short day length—with little direct sunshine and cool to cold temperatures—are not likely to promote rapid growth despite the availability of water. California natives awaken, but not fully. Seeds that germinate after early rains show sluggish growth; bulbs that are ready to sprout new leaves do so very slowly; larger plants such as shrubs and trees may fatten their buds, but little more. Most woody plants continue to slumber. Grasses, however, germinate and grow with vigor, draping the hills in a fine green mantle. Above average rainfall intensifies the green even more. Unseen is the activity below ground, for now is a good time for new roots to start their development, particularly if the soil—as usually is the case—is not frozen. Root growth during this time allows preparation for the "sprint" that plants make when conditions change in spring.

After the New Year, and as spring approaches, days lengthen, gradually at first then ever more rapidly. The sun's energy grows increasingly potent as the angle between it and the earth changes. Daytime temperatures warm and yet the soil remains moist. Now plants are primed for their growth and blossoming.

Most native plants are more than ready. Already-germinated seedlings grow at full speed, and many annual wildflowers now must race against time to produce enough food through photosynthesis in their leaves to carry them through a burst of flowering and seed ripening before moisture is gone. The extent to which they achieve these goals is dependent on the particular pattern of sun and rain for

that year; sometimes a favorable balance between sunshine, warm days, and rains extends over several months; other times, factors converge to create stressful situations. So it is that in favorable years, wildflower displays may be memorable for their abundance, variety, and exuberance; whereas in other years, flower color is meager and quickly fades.

Flowering bulbs—because of their early root growth and stored food—are particularly well adapted to California's spring weather. What affects bulbs most is not the conditions of this spring but rather the conditions that prevailed the previous spring. Bulb-bearing flowers put as much food as possible into next year's bulb (including its flower buds) while also providing food necessary for this year's flowers and seeds. A long, warm, wet spring one year guarantees abundant bloom the next.

Many deciduous trees put out their flowers just as winter has turned the corner to longer days. Then the winds are strong and gusty and the bare branches don't impede the movement of pollen, for most deciduous trees also happen to be wind pollinated. Hazelnuts, alders, willows, oaks, and box elder are examples. Because they rely on strong winds to move pollen—not insects or birds—their flowers have lost their petals. Instead the flowers are often borne in long, dangling chains known as catkins, and the individual flowers are tiny and greenish, brownish, or yellow.

As days warm and pollination is done, deciduous trees open their winter-dormant buds and start the serious business of growing new leaves and twigs. The initial opening of such buds ("bud burst," as it is called) appears rapid because all the parts were already fully formed inside, just waiting for the right conditions. When they finally do open, usually because of warm temperatures, roots take in large quantities of water from moist soils, and all of the growth that results is simply a matter of pumping the new leaves and twigs full of water so they expand.

For evergreen trees, the sequence is similar except that in years with poor rainfall, not all new buds open. In fact, one of the advantages of being evergreen in a summer-droughty climate is you don't have to make a new set of leaves at all; what was there last year will serve for at least a couple of more years before they must be replaced with new leaves.

Many of the trees and shrubs also take care of flowering and seeding in spring. Here there's much more variation than with annual wildflowers, for tree trunks and larger limbs of shrubs are able to store at least some water for later use. And in really bad years, trees and shrubs may flower only scantily, although there's a very interesting phenomenon that may go counter. If there have been several years with poor rainfall, trees or shrubs might suddenly blossom with a vengeance, putting all of their energy and water into making flowers and seeds. This is a last-ditch strategy commonly seen no matter what group of plants we're dealing with, and it somehow represents an anticipation of impending death. What triggers this response is still a mystery.

By spring's end, then, most native plants have—with sufficient water—put on a burst of new growth; have flowered; and have gone to seed. As soils turn dry and temperatures steadily climb, plants prepare to take a rest—this in spite of the fact that now the sun's angle is at its strongest. So one disadvantage of adapting growth to a Mediterranean climate is that plants miss out on one-half of the year's favorably long days because of drought.

Of course remember that coastal California generally does not become hot in summer. The cooling fogs in fact do keep many shrubs and trees flowering longer, but despite extra fog drip, soils still lose water faster than it's being replaced.

Several annual wildflowers begin active growth at the beginning of summer; these plants actually delay flowering and seeding until mid- to late summer when everything appears bone dry. How these fragile-looking plants achieve such success is not well understood; certainly, they have special features that help. Such features include varnishlike substances that are plastered over the leaves and stems to keep them from drying out; taproots and stems that store as much water as possible; and strong "volatile oils" that evaporate on warm days to cool the plants and that taste and smell badly enough to protect vulnerable foliage from hungry animals.

Look for these "miracle" annuals in some of the least favorable spots—places where the grasses are short, along paths where the soil is hard-beaten, or along roadsides. Some of them actually come from other lands and so are not native here: these include the sky-blue-flowered chicory, various dandelions, wild lettuce, and Queen Anne's lace (really a wild form of carrot). Native annuals include the sprawling silvery mats of turkey mullein (leaves lemony-odored when crushed); the miniature blue, curled flowers of turpentine weed; the spiny flower heads of skunk weed; and several tarweeds. The latter bear daisylike flowers with bright yellow rays and centers, but the whole plant is covered with sticky glands that have a tarlike feel and also a strong odor that some people (but not everyone) find pleasant. All of these annuals are "doing their thing" in summer for only one fundamental reason: there is little competition for light and soil nutrients with other plants.

Meanwhile, another important event is happening with many shrubs and some trees. Although these almost always flower before summer, the seeds may not fully ripen until summer or even fall. This is especially true for fruits that take a long while to reach maturity: for example, hazelnuts, buckeyes, and oak acorns, with their rich store of food; or the myriad berries including huckleberry, blackberry, thimbleberry, and elderberry. For berries, it's a question of building up high sugar content and lots of water; for the others, it's building up massive amounts of stored food.

A third category of fruits borne by trees and shrubs also slowly develops all summer long. These include many wind dispersed seeds of trees, although willows are one exception: their seed development is swift following late winter flowering, and the tufts of cotton-covered seeds are blown about while spring winds are still blustery. But ashes and maples—with their winged samaras—and alders—with their tiny cones that bear minute, winged seeds—all wait until late fall to complete the development of their fruits and seeds. Certainly, the reason for the delay between flowering and fruiting is not that they require lots of water or food to complete the process; rather, it's probably the fact that when fall days grow short, the winds grow gusty and strong again, and this is a good time to send seeds on their way.

Fall is a time of slow, lazy changes. Perhaps the single most noticeable event is leaf fall, for as days continue without new moisture, the stress of water loss increases. Since most plants lose water through evaporation from leaf surfaces, shedding leaves makes sense. Poison oak and buckeye are two plants that may lose their leaves as early as mid-summer if drought stresses are unusually severe.

One of the events accompanying leaf fall of course is the dramatic color changes that everyone associates with the "real seasons." The reason this occurs more in the eastern states has to do with the cold nights in fall that destroy the green chlorophyll in leaves before the leaves are shed. Most leaves contain several accessory pigments—often yellow or orange carotenoids, red xanthophylls, or purple anthocyanins. These show clearly when the chlorophyll is missing. Because California's autumn nights may not have the same "bite," many of our deciduous plants' leaves fall without the chlorophyll being destroyed first. Nonetheless if you look carefully, you'll find that certain plants do give color: the most notorious and obvious is poison oak with its bright red autumn leaves.

Most plants continue their summer dormancy right up until the first rains. One of the most unpredictable things about California's autumns is when those first rains are going to come, and then, how often. One or two minor storms really don't get things moving again. When the first substantial rains arrive determines the mushroom season, the reviving of mosses and lichens, and the first flush of new fronds on certain ferns. Three of California's most abundant ferns—California polypody, licorice fern, and California maidenhair—disappear during the drought; that is, their fronds die and the food is temporarily stored under ground in the roots. These ferns signal the return of rains by making a bold new flush of delicate green fronds. But in the case at least of licorice fern there is some question whether it's the moist soil following rains or the day length which determines this renewed growth, for if rains are late you'll often see new fronds struggling to unfurl anyway. Plants may decide when to initiate growth, flowering, or fruiting according to more than one kind of environmental cue: sometimes it's day length, sometimes temperature cycles, and sometimes moisture. And sometimes, they have to all act in concert. Each kind of plant is different, and only by close observation do we come to understand what is important. In the case of licorice fern, day length must be reliable enough most years to assure that rains return, or this fern would have eventually died out.

As rains finally begin, a number of events unfold—different for different sorts of plants—for about the same time the short days have brought back colder temperatures. Most plants with their closest relatives geared to the cold winters of the temperate United States now shut down; this applies to maples, willows, buckeyes, ashes, cottonwoods, and other woody plants that lose their leaves in winter; it also applies to many smaller plants that die back to sturdy underground roots. Such plants include trilliums, star flower, false Solomon's seals, fairy bells, California fuchsia, and others, many of which live in the shade of thick forests.

On the other hand, rains begin the active growth of many nonflowering plants; most of these—except for the conifers—reproduce by spores and are very vulnerable to drying out. The drought, however, doesn't kill them, they simply remain in "suspended animation."

Our seasonal journey has taken us from the first winter rains to the longer, warmer spring days, and from the drying process at spring's end through the long droughty summer and early fall to rejuvenation by the first rains. Each season features a different set of plants or plant parts and therefore holds its own fascination. Nature's changing ways work their magic here, but in a drama very different from the abrupt seasons New Englanders experience.

*Linda Yamane*

We're forever looking at our watches, setting our alarms to get up at a certain "time," scurrying to get to an appointment at a certain "time," to school "on time," or to work "on time." It's a way of life we accept as "normal." It's what we've always known. Likewise, we've grown up with calendars on the desk or wall—neat squares of paper, the days and months of our lives, which we dutifully flip one after

the other or tear off and toss away. Our calendars tell us when one season ends and another begins. First day of summer? Why, it's unseasonally cool for summer, not what summer is "supposed" to be like. Middle of winter? Why, it's as warm as a spring day. This isn't what winter is "supposed" to be like, and we feel confounded that the weather doesn't match our expectations.

In the old times, Native Californians weren't ruled by clocks and calendars. Rather, the passage of time and seasons was indicated by the varying rhythms of the natural world. The days get progressively longer or shorter, colder or warmer, the plants and animals responding accordingly, developing and completing their cycles. People living close to and depending so directly upon their immediate environment are finely attuned to the subtle changes that signal the impending readiness of those resources upon which they depend. For California Indian peoples, seasons might correspond to the maturing of acorns, bulbs, grass seeds, or berries. There are seaweed and shellfish harvestings, and the ever-important salmon runs.

Time was told by linked events. A festival was held to celebrate the ripening of strawberries, and people knew that within so many days the brodiaea bulbs would be especially ripe for harvesting. When a certain bird first sang, the Hupa knew that the salmon would soon be running. When the elderberry flowers bloomed, the Kashaya Pomo no longer went to the seashore to gather shellfish, but waited until the berries ripened. These calendars varied, each community attuned to its own environment.

In contrast to our four calendar seasons of spring, summer, fall, and winter, the desert Cahuilla observed eight seasons, each relating to the growth cycle of mesquite, an important food plant. These were: budding of trees, blossoming of trees, trees forming seed pods, seed pod ripening time, falling of seed pods, midsummer, cool days and cold days. The Maidu distinguished the seasons of flower time, dust time, seed time, and snow time, while Coast Miwok seasons included a ground coming out season, a hot season, a short day season, and a fourth that has since been forgotten. Root time, fire gone time, hot day time, and leaf on top time are the four seasons reportedly observed by the Cahto.

We live a very different kind of life today, and it is fundamentally so different that it is difficult, if not impossible, for us to truly imagine the reality of living the other way. We do not live in close contact with the natural world and so are not keenly aware of the many nuances of nature that signal the changing seasons. We do, however, have our own unique seasons to which many modern people are equally finely attuned. They are called baseball season, football season, basketball season, hockey season, ski season, and tourist season—to name a few.

There is so much about modern culture that seems to diminish seasonality. Rather than adjusting ourselves to the changing seasons, we have devised ways to modify our environment—at least our immediate mini-environments. On unbearably hot days, we sit blissfully and gratefully in air-conditioned buildings and cars. In the cold of winter, we turn on our heaters, adjusting our environment to suit us. We seem to have developed a propensity for leveling out the extremes, for seeking the comfort of sameness. While on the one hand our avoidance of discomfort is part of human nature, the predictability we have created has also taken away our need to adapt. We are not content to be partners with the rest of earth's beings, receiving the seasonal differences our world provides. We have taken a position of control over the seasons, even to the point of making "snow" when there is not "enough" of it on the ski slopes.

We seem to want all things at all times. We want to swim in winter as well as summer, so we make heated and indoor swimming pools. Irrigation creates a year-round "rainy season," making it possible to grow plants in places they wouldn't normally grow and in seasons they wouldn't normally survive. Whereas we can buy lettuce at any time of the year, native peoples only had fresh greens in certain seasons. There are stories of people lying in fields, devouring clover leaves in revelry by the handfuls, and special dances celebrated the appearance of clover in the spring. Because we rarely have to go without the foods we want, I doubt that we experience anything close to the joy and appreciation of the past. Surely we've noticed, though, how especially satisfying and flavorful food is when we are exceptionally hungry. Imagine the richness and value of stories told only on winter nights. Perhaps by providing so much for ourselves we have robbed ourselves of the discipline of waiting and the subsequent joy and reverence of receiving.

# Gathering

*Linda Yamane*

Today we typically respond and relate to what we see around us, the visual having become our normal, almost exclusive, interaction with the plant world. We are taught that it is bad to pick plants, even to pluck a leaf—we are not to disturb nature.

Perhaps this philosophy is a protective reaction to the long period of wanton destruction of natural places. Plant and animal habitats continue to disappear as our population continues to grow, and so public land agencies have been developed, in some cases to preserve wild places, in others to manage natural resources for certain specified or marketable ends.

Unfortunately for Native peoples, the values represented in determining public land usage have not historically incorporated traditional Native perspectives. Indian people have long held close relationships with the plants. Plants have provided food, shelter, clothing, baskets, cordage, games, musical instruments, hunting and other tools, medicines—the list could go on. Plants were, and still are, thanked with offerings and sung to. Plants are cared for—bulbs and plantlets are replanted, seeds dispersed.

Traditional Indian land management practices such as pruning, coppicing, and burning actually benefit the plants. These traditional horticultural techniques control the spread of disease, stimulate vigorous growth and greater fruit production, relieve plants of dead or dying limbs, and have been shown in field experiments to enhance plant resources, not diminish them. The harvesting of underground roots and tubers involves digging, which loosens and aerates the soil, enhancing growth potential as well.

Seasonal burning is known to control the growth of brush and promote the growth of seed-producing grasses and herbaceous plants. In Native California, burning was most commonly done in the fall, after the edible seed crops had been harvested and the remaining seeds scattered to the ground. Modern experiments have proven that the greatest number of new seedlings sprout in spring after a fall burning. In fact, certain plants only appear in substantial quantities after a fire. Burning also provides feed for game animals, and experiments in areas of prescribed burns have shown dramatic increases in deer population.

Fire also kills parasites that infest plants and render them unhealthy. Burning off the duff in oak woodlands, for example, not only kept the areas free of encroaching brush, making it easier to harvest the acorn crop; it killed the acorn weevil as well.

The regular, controlled burns of pre-European times were responsible for the vast park-like landscapes that greeted the early European visitors. After 1900, when burning was against the law, forests and woodlands became thick with brush, and when an area is not regularly burned a fire is more likely to be catastrophic when it does occur. Dense undergrowth provides fuel that burns to such intensity that it spreads to the tops of trees, devastating huge areas.

Burning was also important to California Indian basketweavers. A number of basketry plants—such as hazel, maple, redbud, and sourberry—yield long, straight, uninfested shoots when burned. Bear grass and deergrass produce the best basketry materials when burned. Bear grass that has not been burned will not be flexible enough for proper weaving. In the absence of burning, pruning will also stimulate the growth of long, straight shoots. Deergrass that is neither pruned nor burned will eventually waste away in its own dead material. In a modern society that prohibits the burning of land, contemporary weavers find it difficult to carry on their weaving traditions without the quality basketry materials they so desperately need.

While the philosophy of non-use is intended to benefit the plants, this policy can sometimes result in their eventual deterioration and even demise. Gardeners and agriculturists in vineyard and orchard know the positive effects that pruning, cultivation, and thinning have on the health and productivity of plants. Thoughtful and limited use of plants actually sustains their health and vitality. Yet modern land management policies have not included indigenous management techniques, nor have traditional plant gathering practices been considered legitimate land use. This reflects differences in cultural values and perspectives.

Today nearly every public land management agency has initiated dialogue or designed programs to develop partnerships with or incorporate input from

We are taught that it is bad to pick plants, even to pluck a leaf—

DO NOT TOUCH

We need access to the plants in order to carry on our traditions,

and the plants need us as much as we need them.

Indian peoples. Native people—basketweavers and gatherers—are telling the same story everywhere. We need access to the plants in order to carry on our traditions, and the plants need us as much as we need them. Some agencies are beginning to appreciate traditional needs and are responding with some limited controlled burns. As a result, basketweavers don't have to struggle quite so hard to carry on their traditions, and Native Americans are beginning to be recognized as legitimate natural resource managers.

# Crops & Other Aliens

*Glenn Keator*

California is a land rich in naturally occurring plants that nurture humanity in many respects: with food, fiber, drink, lumber, medicine, and much more. This native heritage has been mainly neglected or forgotten, replaced with a wide variety of alien plants whose purposes are similar. Their homelands represent such far-flung places as the rainforests of Central America, the highlands of South America, the uplands of Mexico, the Mediterranean basin, the steppes of the Middle East, and the mountains of China and India.

Some of these plants are adapted to climates similar to ours, and flourish with a regime of mild, wet winters and hot, dry summers, without much extra help. Such

"When you see something that's just about everywhere and there's lots of them, there's a meaning to that....That's the way with some of these things we see here; there's a meaning to all of it, but it just doesn't come to you, or if they teach it to you in the schools, that doesn't mean you know it, just means you learned it. You got to live right to know the meaning to these things."

—Laura Somersal,
Pomo/Alexander Valley Wappo

crops as olives, citruses, cork, artichokes, and grapes are bound to flourish in California's fertile valley soils without making undue demands upon the environment. But many others—perhaps the majority—demand richer soils and plentiful, seemingly boundless amounts of water.

Yet such interlopers—the lettuces, asparagus, fruit trees, cotton, date palms, avocados, and a wide variety of vegetables—find conditions that are easily modified to meet their demands through the practices of agribusiness. Agribusiness has provided the possibility—on a large scale—of growing vast numbers of valuable crops, where once there was little but dry desert and semidesert conditions. The massive wholesale movement of waters from California's mountain snowfields and aquifers has allowed the watering of previously arid lands. Such lands often have soils that relinquish their fertility to vigorous crops when their bonds with aridity have been broken.

Agribusiness combines the ability to move waters and use them in distant places with sophisticated machinery that harvests single crops on a vast scale, making the everyday price of needed foods and staples practical to the average consumer. Because such monocultures are vulnerable, however, to the ravages of predation by insects and infections by bacterial and fungal diseases, the necessary antidote has been the invention of a large pharmacopia of chemicals aimed toward effective control. Such chemicals allow the efficient growth of near disease-free plants within these monocultures. Other chemicals have been developed,

that when spread wholesale on soils, enrich them and fertilize them for even greater production.

The problem with this paradisacal view of efficient and bountiful agriculture is that what has evolved has been a whole new set of problems that are exacerbated by the solutions involved. We are now faced with either solving these problems through new technology or reverting to more "primitive" and less "efficient" methods of growing crops. There is the possibility that we need to return to the culture of crops that are solely adapted to our soils and climate such as they are in their natural condition.

*Linda Yamane*

The introduction of plants and animals alien to California changed the landscape in ways that had long-lasting repercussions for Native peoples, and I can't help but feel sad for what has been lost. I will never see the bunch grasses that disappeared, outmatched by cattle and European annuals, or the meadows filled with seed plants once central to California Indian life. In my Rumsien Ohlone language there is a word "o-chons"—that translates as "seed times." I can only begin to imagine the full meaning of that word.

There are written accounts of early encounters between Native Californians and European travelers describing the flavorful porridges and seed cakes presented to the first European visitors. There are other accounts, from Ohlone people in the early part of this century, describing how cattle ate the seeds, leaving few to sprout.

In the old times, they reminisced, there was no such thing as a bad year. In some places orchards were planted on land where favorite seed flowers grew. For a while flowers persisted, but the orchards were plowed before the plants had a chance to seed. Now I search for a special flower so I can make the little seed cakes like before—but in my life I have not yet seen one.

Native people, resourceful as always, adapted and did what they could with the new situation. How well I remember my grandmother's passion for mustard greens. And some Ohlone women substituted seeds from certain introduced plants for their "pinoles" and "atoles," or learned to use new plant remedies. When up in years, my grandma always took her black bag, filled with an assortment of medicines, with her when she went to doctor her many sisters and brothers. Inside were modern remedies like aspirin and "Vicks," alongside little wrapped bundles of medicinal herbs, both native and new. So far as I could tell she considered them all to be equally useful. She was not one to lament—she accepted what was her way of life.

But forgive me if I feel twinges of resentment when I see plants that don't belong. I know that eucalyptus didn't ask to be brought here, and as a child I enjoyed its towering presence, bold fragrance, and wonderful ribbons of bark. Today, however, such plants mainly remind me of unwelcome change, hardship, and loss.

# Mushrooms

*Glenn Keator*

With the return of the fall rains, mushrooms pop up as if by magic overnight. The minute we see one, we're likely to recognize it, for most mushrooms bear some kind of resemblance to the store mushrooms we love to cook with.

Although the old textbooks call mushrooms plants, there's increasing evidence that they belong to a kingdom all by themselves. The old guessing game in which everything can be categorized as "animal, vegetable, or mineral" doesn't apply anymore. Instead, the living realm is divided into bacteria (Monera), single-celled organisms (Protista), mushrooms (Fungi), plants, and animals. This new way of looking at things is an attempt to overcome inconsistencies between animals and plants, but the truth is that no matter how we try to create self-inclusive categories, Nature always has exceptions and surprises. Nonetheless the new system sorts things out a bit better, giving us reason to believe that evolution has proceeded in more than the classical two directions (plants and animals).

Mushrooms or fungi represent their own special line (or more accurately, lines) of evolution, possibly from a primitive animal-like ancestor. Even though their growth pattern, their sedentary nature, and their reproduction by spores (single-celled microscopic structures considered typical of plants) create the illusion that fungi are plants, new analyses of their genetic material (DNA) indicate a closer relationship to animal groups. Animal-like characteristics include the substance chitin (chemically related to starch and common in the exoskeletons of insects and other "primitive" animals) and their overall biological activity. Many of the poisons that affect animal life also affect fungi; it has been suggested that this is the reason so many fungal diseases (ring worm, athlete's foot, and more serious internal diseases such as aspergillosis) are resistant to treatment, or the treatments end up being toxic to the patient. Bacteria, on the other hand are considered closely allied to plants, and are generally easier to treat with antibiotics.

What are fungi? Although the group is diverse and contains numerous organisms that are decidedly not mushrooms, the general overall similarity includes usually chitinous cell walls, and a basic feeding body of long, narrow, colorless to white tubes called hyphae. Multiply-branched hyphae create a body called the mycelium from the Greek word "mykos" for mushroom. The minute hyphae rapidly penetrate and spread in whatever they're living on, be they living plants (tissues in leaves, stems, or roots), living animals (tissues in the skin layer or sometimes in the internal organs), or dead and decaying organic materials (such as rotting logs, leaf duff, dung, or dead animals).

Unlike most true plants, fungi are totally devoid of green chlorophyll and thus cannot photosynthesize (make their own food through sunlight, carbon dioxide, and water). Instead they live as "saprophytes" on dead organisms or as "parasites" on live animals or plants. A few even live in special mutualistic relationships, such as the mycorrhizae of plant roots or as lichens.

Whatever the manner of obtaining food, fungi are seldom apparent until they "fruit," reproducing either by asexual spores (spores that grow directly into new plants identical to the parents) or sexual spores (spores that grow into some new or different generation). In fact, fungi have among the most complex life cycles of any known organisms; some of the parasitic rusts and smuts make four or five different kinds of spores!

We notice fungi when they are fruiting, for that is when the spores must come to the surface, where winds can carry them away to new places to grow. In the "lower" fungi—including blights and mildews—spores often cover the leaf or skin surfaces with a powder (millions of spores under magnification), but in the "higher" fungi, spores are produced on highly modified structures that are readily noticed. Such structures are what we call mushrooms or fruiting bodies. We see the carved black-brown fruiting bodies of elf saddles in pine woods; the white "golf balls" of puffballs among grasses; and the white to tan caps, chocolate brown gills, and ringed stalks of agarics in grasses or woods.

Fungi also benefit most plants by the formation of mycorrhizae, quite possibly the most important of mutualistic relationships. In such partnerships, both participants—plant and fungus—benefit.

Mycorrhizae are special soil-inhabiting fungi that also act as saprophytes, their hyphae ramifying through the earth to absorb water after rains and nutrients from dead leaves. These fungal strands also coat the outside of plant roots, sometimes penetrating a short ways, where they receive sugars that have been manufactured by leaves and sent to the roots for nourishment. In return, plant roots absorb some of the recycled nutrients from dead leaves including nitrates, phosphates, and calcium and extra water as well. Fungal hyphae act like root extensions, making the host plant's water and nutrient uptake far more efficient, and in return hyphae have their own nutrient intake supplemented by energy-rich sugars of the plant.

Some mycorrhizae are highly specialized and live with only certain plant roots: these specialized kinds are typical of roots of conifers (cone-bearing trees such as pines, firs, and spruces, for example), orchids, and heathers. Many of these live in highly acid soils, where essential nitrogen-fixing bacteria cannot survive. Nitrates piped through the hyphal network to roots of these plants make up for the direct absorption from soils, allowing these specialized plant groups to thrive where others fail. If you walk through Monterey pine groves in late November, for example, you'll notice the vivid storybook fruits of fly agarics (*Amanitas*) with their rounded candy-apple red caps flecked with white warts. These amanitas are notable mycorrhizal partners with conifers.

Other mycorrhizae serve specific needs of highly specialized flowering plants, such as orchids and wintergreens. These plants have seeds so tiny they have none of the usual food reserves packed around the embryo plantlet inside. Instead, seeds must land next to the right fungus which penetrates and then feeds the new embryo until it can establish itself and start to photosynthesize. Perhaps this relationship first developed as a way of the mycorrhizae later getting sugars from the green leaves of most orchids and wintergreens, but several members of their families have lost their chlorophyll—growing as they do in the deep shade of mature forests—and no longer can live by themselves. Instead they live through these mycorrhizal fungi that are hooked up to another food source, roots of trees. Such specialized flowering plants are incorrectly called saprophytes, for it was once widely believed that their roots broke down leaf litter and reused the nutrients contained by themselves. We see several intriguing examples of these fungal parasites: coralroots (*Coralorrhiza* spp.) and ghost or phantom orchid (*Eburophyton austinae*) for the orchids, and snow plant (*Sarcodes sanguinea*) and pine drops (*Pterospora andromedea*) for the wintergreens.

The fungi we most often notice are those with conspicuous fruiting bodies, loosely known as mushrooms. Any time the word mushroom is mentioned, the alternate word toadstool also comes up. What is the difference? Mushrooms comprise a large group of fungi with varied sorts of fruiting bodies, whereas toadstools usually refer—but in a very general way—to poisonous or harmful mushrooms. A child's fairy tale book often portrays a red toadstool on the cover, one reviled and revered around the northern hemisphere called the fly agaric (*Amanita muscaria*). Why this is the symbol for toadstools is fascinating and instructive. Fly agarics live in the leaf litter of damp forests from Siberia to northern Europe and across North America, occurring south into the high Rockies and along our Pacific coast. In California look for fly agarics under Bishop or Monterey pines after fall rains begin, temperatures turn cool, and days shorten.

Wherever it grows, fly agaric has unseen hyphae that are mycorrhizal with tree roots: birches in Siberia, spruces in the Colorado Rockies, and Monterey and Bishop pines on the California coast. Evidently there are different races of fly agarics according to the tree association they form, but they all look alike when they fruit, and are unmistakable then: broad, deep red caps push through the leaf duff in November and December, wearing white "spots" or patches. The cap expands as it meets the air, and slowly the color fades and the spots wear off. Underneath, the cap is ribbed with many radiating white gills that don't quite connect to the white stem. Look down the stem, and you'll see a white ring about half way down; at the bottom, the stem is enlarged and has several series of partly broken white rings (the volva). All of these features make identification of this kind certain. One more test to make is determining spore color: place the cap with the gill side down on a piece of black paper and leave it overnight. When you lift the cap in the morning, an exact print of the gills is left behind by the thousands of tiny spores that fall on to the paper. For the fly agaric, they'll be white.

Why does the fly agaric have such fascination beyond being a striking mushroom important to certain trees? Well, there's a long legend that goes with it, starting at the dawn of history. These mushrooms have been sacred to many Siberian peoples whose shamans go into trance and have visions from eating the cap. Several scholars believe that mystical beginnings of some religions are a result of these visions. Remember that in that far northern region, the fly agaric accompanies birches. The same mushroom in California seems downright dangerous and unpleasant to use: although no deaths are authenticated from eating the caps, people get very ill, and with few of the visions so well documented from the Old World. It seems logical, therefore, that our race of agarics is different chemically, and it is not unlikely that this difference is due in part to its mycorrhizal partner. Hence, those fairy tale books are possibly indicating two things: the seemingly magical quality of this mushroom and also its potential for harm. Some of the sister species—death angel (*Amanita velosa*), for example—are lethal when eaten, while others such as the coccora (*Amanita caesaria*) are considered edible and choice.

Of course the fly agaric is but one of hundreds of kinds of mushrooms we find in our fields and woods. Most of them

fruit after the fall rains begin, with a peak between November and January. Since flowering plants have mostly gone dormant then, it's great fun to go mushrooming, even if you don't intend to eat any. In fact, you should never eat a mushroom unless you're fully acquainted with its identity and can tell it from any possible harmful look-alikes. There is no magic formula to tell toadstools from the edible kinds.

Aside from their edibility or poisonous qualities, mushrooms are fascinating and beautiful in and by themselves. Look for the fragile, pale yellow, bell-like caps of *Bolbitius* growing on old dung in grasses; the tiers of multi-hued turkey tails and other shelf fungi on old logs; the orange blobs and dabs of gelatinous witch's butter on twigs; the tiny white balls set in a holder of a black, five-pointed star of earth stars on barren soils; and the tiny white-gray digits of deadman's fingers on decaying wood.

As you begin to search, you'll soon find a whole undiscovered world of new forms. There are little brown mushrooms, big brown mushrooms, medium-sized brown mushrooms, and then the really interesting ones, such as those in bright colors. Just about every color imaginable occurs, from the slimy yellow and orange caps of *Hygrocybe* to the blues and violets of blewitts, the deep rose, pink, or dark reds and tans of *Russula,* and the subdued greens and tans of parrot mushroom. Caps may measure only a few millimeters across or extend to well over a foot; they may be flat, steeply cone-shaped, rounded, or

dimpled in the middle. The surface of the cap may be smooth and dry, gelatinous and sticky, scaly, or hairy, with blotches, spots, or patches of universal veil or not. The gills underneath may be widely spaced, closely crowded, forked or one piece, wavy or straight, thick and waxy, or thin and dry. Gills may touch the stem, run down the stem, or stand away from the stem, leaving a space between. The stem may be stout and solid, thin and hollow, fragile or firm and leathery, ringed or not. The list of variables goes on and on.

Then there are the mushrooms without gills. Some mushrooms bear their spores on teeth extended below the cap (the tooth fungi); some bear them on veinlike folds (the chanterelles); some on coral-like single or complex branches (the coral fungi); yet others produce their spores inside minute hollow tubes. Tubed (or pore-bearing mushrooms) belong to two important groups: the boletes, with a soft cap that resembles an ordinary gilled mushroom from above; and the shelf fungi or polypores, with tough leathery to woody caps that usually grow as brackets along dying tree trunks or on logs. Among these polypores, one kind—the turkey tails—has wide bands of varying colors on the top of the cap and another kind—the artist's conch—makes huge woodlike caps whose undersurface, where the pores are located, may be marked. Conches are particularly striking and important to forest ecology, for their hyphae often begin the process of weakening old trees by feeding on the inner living bark. Only later as the tree is dead or dying do the fruiting bodies appear, and then they grow for many years, each year adding a new half ring of tissue much in the way trees add a new ring of wood each year.

Finally, there are the mushrooms with spores borne inside the fruiting body: puffballs, earth stars, stink horns, and birdsnest fungi. Each is fascinating in its own way: puffballs develop a hole at the top and spores are blown out by bellows action of the fruiting body which responds to changes in humidity; stink horns develop their spores within a stinking, slimy mass to attract dung beetles and flies for the purposes of dissemination; birdsnest fungi make their spores inside egg cases, the eggs lying at the bottom of a perfect miniature nest. The force of raindrops knocks these eggs from their nest, and the sticky threads attached to the eggs wind around a nearby branch, anchoring them to a new home.

Were it not for the way in which fungi gain a living as saprophytes or parasites, the living world would be a very different place. Although we lament the action of rusts and smuts on many of our major crops, or mildews and blights on the vigor of garden flowers and other food plants, or the food spoilage caused by molds on bread and jams, fungi are indispensable to the earth's ecosystems: saprophytic fungi (along with bacteria) are the only organisms that ultimately recycle dead plants and animals by breaking them down into their component parts. Thus in the complex food web of ecosystems, fungi serve as the "decomposers," without which the earth would be choked under billions of tons

of organic refuse. This recycling process means that leaves become leaf skeletons, that old stumps and logs are reduced to mulch, and that compost piles decompose until they become useable and important humus for enriching garden soils. Were it not for decomposers, grassy meadows would soon be buried under a vast mulch of dead leaves and animals, and marshes and ponds would fill to overflowing with animal carcasses and plant debris.

# Lichens

*Glenn Keator*

There is probably no other group of "plants" so widespread yet so mysterious and poorly known as lichens. Lichens occur in all the places other plants do not: raw rock faces from hot deserts to the salt spray zone along the coast; bark and twigs of trees and shrubs; bare earth in the tundra, on sand dunes, or above timberline in high mountains; and rocky banks everywhere. Many books call them "ecological pioneers," meaning they're the first to inhabit newly created sites such as lava flows, newly exposed rock surfaces, or roadcuts. In coastal regions look for them along rocky banks, cliffs, and even on barren soil.

Despite what's been said about them, it's unlikely lichen contribute significantly to the formation of new soils through the breakdown of rocks: they grow much too slowly. The real forces creating new

soils are those geological processes called weathering: battering rocks through heavy rains, sand and dust storms, and freezing and thawing.

Lichens are really two organisms rolled into one: fungus and alga. As discussed above, fungi are in a kingdom separate from true plants, while the algae may or may not be plants according to broad or narrow definitions. The fungus part creates a densely interlaced web of strands (hyphae) which to the naked eye looks like a single plant. Buried beneath these strands, often in the upper half, is a band of microscopic single-celled or filamentous algae, living within the protection of these strands where they're prevented from drying out and where they receive enough sunlight to photosynthesize. The fungal body anchors the lichen to whatever it's growing on—soil, rock, or bark—and absorbs moisture from humid air or from rains. It also absorbs dissolved minerals from its anchoring filaments. The algal cells, being green and containing chlorophyll, are able to combine carbon dioxide from the air with water, turning them into usable sugars, which both partners use for energy or convert into new materials for growth.

For years it has been debated whether the relationship between partners is equal or lopsided, and many textbooks cite lichens as the perfect "mutualistic relationship." But research is uncovering fascinating aspects: the fungus cannot grow on its own without the help of the algae, having evidently lost its ability to obtain food otherwise. The algal partner by contrast can live outside the lichen

because it creates its own food, but could not possibly survive under the harsh conditions where most lichens grow—it would simply dry out too fast.

Nevertheless, research in the natural world has shown that seldom does the algal partner exist outside its lichen home, even when conditions are optimum. There are exceptions. Certainly there are microscopic "free living" algae on oak branches along our immediate coast, where high humidity occurs. Conceivably some of these algae may still occasionally find an acceptacle fungal partner to create a new lichen, but this event is probably rare, and most lichens reproduce themselves from already existing lichens.

In fact, lichen reproduction is another mysterious aspect of our story, for the fungal partner still goes through the energy-wasteful step of creating new fungal spores in special fruiting bodies called apothecia. These apothecia are minute cups or discs that sit on the upper surface of the lichen; lining the bottom of these cups are innumerable microscopic sacs that explode as they dry and send out clouds of spores. Since only the fungus is involved, these spores must not only find a new home, but a home with exactly the right algal partner. The likelihood of any fungal spore finding a proper home is already exceedingly small, but the added requirement of meeting the "other half" makes the continued survival and development of any spore highly unlikely! The question remains whether there is any benefit to the lichen to continue to make spores. Do they succeed often enough to justify the expenditure of energy? Or do they simply carry on doing this because they've never lost their genetic program to do so?

How then do lichens succeed in reproducing themselves if not this way? Many lichens simply fragment when they're dry: try feeling a dry lichen, and you'll find that it's very brittle and easily broken. If the severed fragment finds a suitable new home, it starts to grow on its own as soon as adequate water is available. Lichens may also produce tiny flourlike beads called soredia on their upper surface; each bead is a ball of entwined fungal threads enclosing several algal cells. And there are other specialized structures on certain lichens which contain fungus and alga and also act as vehicles of lichen reproduction.

Lichens occur in bewildering variety and form, but the most basic classification divides them into these categories:

*Crustose lichens.* These tiny lichens form a super-dense crust on rock surfaces—sometimes also tree bark—where they're so tightly cemented to the rock that the two are inseparable. Look for bare rock surfaces that seem painted with a variety of colors: sometimes drab grays, whites, and black; other times kaleidoscopic mottling of yellows, greens,

reds, and oranges—occasionally even blue. One particularly brilliant, orange-red lichen grows only where marine birds excrete uric acid; others are much less particular.

*Foliose lichens.* These lichens are flattened ribbons that are lobed; they're often loosely attached so that they may be peeled from the rocks, soil, or bark they're growing on. The underside may be quite differently colored from the upperside, and there are often tiny threads on the lower surface that attach the lichen to bark or rock. Many beginning naturalists mistake the gray-green or dull green foliose lichens for liverworts, but generally the color of the latter is deeper green, and a hand lens will reveal tiny pores evenly sprinkled on the surface of these liverworts.

*Fruticose lichens.* These lichens resemble miniature shrubs or trees, or form long intertwined strands like lace or cheesecloth. Perhaps our largest, most conspicuous lichens belong here: old man's beard (*Usnea* spp.) is a grayish-green lichen on tree branches looking all the world like a beard; lace lichen (often erroneously called Spanish moss, *Ramalina*) creates long, delicate networks of gray-green lichens hung from tree branches.

Finally, lichens are useful in strange and seldom thought-of ways. Many are extremely accurate indicators of air pollution; in fact, lichens are missing or sparse on bark of trees in city parks and along streets. Some lichens have pigments that change color according to the pH (acidity or alkalinity) of their surroundings; these pigments are used to

make litmus paper. Other lichens have the algal partner technically as blue-green bacteria (formerly called blue-green algae, hence the confusion): these specialized bacteria can "fix" nitrogen in a form usable for higher plants. Lichens also occasionally serve as emergency food; it's doubtful that humans would eat them by choice, for most are very fibrous and unappetizing. Nonetheless several northern lichens have been used for food during times of famine, as during long winters when little other vegetable matter is available. Reindeer are considerably more dependent on lichens, and the famous reindeer "moss" of northern tundras is really a multi-branched lichen that is a major ingredient in the reindeer diet.

Lichens remain little more than strange growths and excrescences to the majority of naturalists, yet the mysteries contained within and the beauties displayed under the lens are reason to explore their intricacies and contemplate their wondrous role.

# Seaweed & Algae

*Glenn Keator*

Some of our least known plants, the algae, live mostly in water. Many are so tiny—with single or small clumps of cells—that a powerful microscope is necessary to see them. Few live on land, and those that do live only along the coast where summer fogs prevail and humidity is high all year. The commonly used term "seaweed" applies to algae that live in the ocean; sometimes, you'll hear the term "pond scum" applied to certain freshwater kinds.

Algae are the most basic of plants. So simple are the microscopic kinds—single cells that move around under their own power—that they have been combined with the simplest animals, the protozoa, in their kingdom known as Protista. The reason? Simple algae straddle the line which supposedly separates animals from plants. For example, animals can move around under their own power, whereas plants are thought of as fixed in place. Yet some single-celled algae have whiplike flagellae by which they move. Animals have no cellulose cell walls around their cells; plants supposedly do. Yet several algae lack cell walls. Animals cannot make their own food by photosynthesis while green plants can. Yet under some circumstances certain algae may lack chlorophyll—the green pigment necessary for photosynthesis. And so it goes; for every trait that separates animals from plants, there are exceptions.

The creation of a new kingdom, such as Protista, makes it seem as if we've solved the classification problem. But look closely and more problems arise. For protists are supposed to be single cells, simple filaments one cell thick, or colonies; yet the line between a simple filament as opposed to complexly branched filaments is arbitrary, and once again we have overlapping groups. Nor do the algae belong to one great group; rather, they're so varied that botanists split them into several different divisions—the equivalent of phyla used to separate major animal groups.

The reasons that algae are considered to fall into several very distinctive groups are based on fundamental properties of metabolism; in other words, how algae live at the most basic level. Each algal division has different ways of storing food for later use. Each algal division has different kinds of chlorophyll—those green pigments so necessary for trapping light energy in the process of photosynthesis—and different accessory pigments (pigments of other colors which also trap light energy). And finally, each algal division has its own ways of reproducing; life cycles among the algae are the most complex and varied of the different plant groups.

By comparison, all the land plants—from mosses to the most specialized flowering plants—have the same basic stored food (insoluble starch), the same two kinds of chlorophyll (chlorophylls A and B), many of the same accessory pigments (primarily xanthophylls and carotenes), and similar life cycles. This means that the algae taken as a whole are more varied in the very basics of how they live and how they reproduce than all green land plants. This is probably so because of their much greater antiquity, although we have little solid fossil evidence to tell just how old algal groups really are!

What are some of these algal divisions? Can they be identified by the average interested naturalist? Let's examine the three major divisions (but by no means the only important ones). These comprise the green, brown, and red algae.

*Green algae.* The greens are a large group adapted to both fresh and salt water. They also have the largest range of any algal division, for they vary from single-celled to colonial, filamentous, bladelike, or complex conglomerations of intertwined filaments. Accordingly, they may be microscopic (but produced by the millions) or easily seen. They are all held together by the possession of chlorophylls A and B, carotenes as accessory pigments, starch as stored food, and clearly marked alternation of generations in their life cycles. Reproductive cells can swim on their own. It is these features that cause botanists to say that all green land plants had green algae as ancestors!

What do green algae look like? Freshwater greens that are easily seen are long, slimy, bright green strands growing in quiet streams, ponds, or lake margins. If you're visiting tidepools, look for a flattened leaflike alga called sea lettuce (genus *Ulva*) or dense, matted cushions called sea moss (genus *Cladophora*). Both of the latter are vivid green.

Rather less obvious green algae include the near-black, fingerlike spongy colonies called deadman's fingers (genus *Codium*) found in tide pools, or the brilliant, red-orange spongy masses of *Trentepohlia* that paint coastal rocks and twigs of cypresses. Both are perfect examples of green algae that don't appear green to the human eye. In the case of deadman's fingers, the strands that make up the fingers are so densely intertwined and have so much chlorophyll that they appear almost black. In the case of *Trentepohlia*, there is a special hemoglobinlike pigment which

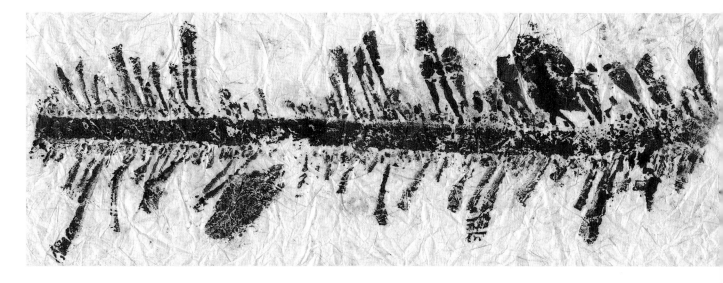

occurs in such quantity that it masks the green chlorophyll.

*Brown algae.* Our second group, the brown algae, have chlorophylls A and C, fucoxanthin as accessory pigment (brown color), a complex sugar as stored food, and also alternation of generations with at least some reproductive cells motile. It's the fucoxanthin pigments that give most brown algae their brownish color, and may partially hide the green chlorophyll. Browns also come in a wide range of forms, although few are truly microscopic. Most browns live in saltwater and are notable and obvious components of tidal zones and tide pools. Most are quite noticeable to the casual visitor. In fact, a large proportion of the seaweeds are forms of brown algae.

Some browns, such as *Leathesia*, grow as brown patches on tide pool rocks. If you didn't know better, you'd think that these patches are part of the rock, or even that they're crusts of thick oil from oil spills!

Larger browns include the various rockweeds, which are abundant in the upper zones of tide pools. Rockweeds usually are flattened, with repeatedly forking branches, each branch a mirror image of its partner. When they're fertile, these branches bear hollow, inflated sacs containing the eggs or sperms. Rockweeds are among our most conspicuous seaweeds.

The largest browns are known as kelps. Kelps are important and interesting for many reasons: they are the largest of all saltwater plants (giant kelps may grow to over 100 feet long!); they have complex structures comparable to the roots, stems, and leaves of land plants; they grow faster than most other plants (except perhaps such wonders as cottonwood trees and bamboo); and they have many uses in present-day manufacturing. Kelps range from relatively "small" plants—up to two or three feet long from tide pools—to giants growing beyond the normal tide line at depths of up to 50 or more feet.

Typical kelps have a multibranched "holdfast" that grips tightly to rocks to anchor them (they often die when these holdfasts are destroyed), long cablelike stipes (the equivalent of a stem in land plants), and flattened blades that function as leaves for photosynthesis. The stipes are even more remarkable when opened, for they contain a core of food-conducting cells much like those found in land plants. And the blades found at the tips of stipes where they reach the water's surface for abundant light are not only flattened and leaflike, they also are buoyed by one to several hollow bladders so that they stay on the surface.

Two distinctive kelps along our coast are the sea palms, *Postelsia*, that look like short, brown palm trees and have flexible stipes to take the pounding of the surf since they live right at the cusp of the tidal zone; and giant kelp, whose immense stipes reach 50 to 100 feet in one season's growth, and are tipped by one enormous float or bladder (the ones everyone likes to pop when walking along the beach). The giant kelps grow in thick beds normally well below the low tide line and create the equivalent of underwater forests.

These forests are important for a number of reasons, the most wonderful of which is the creation of myriad niches for specialized plants and animals, just as trees do on land. For example, there are many algae that grow only on the stipes or blades of these kelps, and there are many fish that hide behind the stipes or even live within the contorted confines of the holdfasts. For us, giant kelp are important commercial sources of alginates, substances extracted that act as emulsifiers. Emulsifiers are those seemingly magical substances that allow usually unmixable ingredients to stay mixed. They're used to smooth paints, toothpaste, salad dressings, and ice cream, to name a few common uses.

Perhaps no place is the alternation of generations more clear-cut nor more dramatic than with the kelps. The microscopic spores produced by the kelp blades must sink to the bottom of the ocean; only then do they start to grow. But instead of growing directly into a new giant kelp complete with holdfast, stipe, and blade they grow instead into a minute plantlet on the ocean's bottom. This plantlet is so small it's barely visible to the naked eye. Its only job is to

produce eggs and sperms. When a sperm has found an egg and fertilized it, only then can the new kelp plant begin to grow.

*Red algae.* Reds are trickiest to identify, for their plant body assumes a multitude of forms, most of them large enough to be seen without magnification. Like the browns, most reds live in the sea, largely in tidepools. Reds are distinguished by chlorophylls A and E, accessory pigments called phycocyanins (bluish and pinkish pigments), stored food as floridean starch (chemically different from ordinary starch), and reproduction often involving three generations. None of the cells are motile; instead spores are moved around by wave action.

Red algae differ so much from other algal groups that their origins are unknown. Their distinctive pigmentation is only found in one other group: the blue-green bacteria. In fact, so algalike are the blue-greens, that they were once incorrectly classified as algae! Their fundamental cell structure, lack of sexual reproduction, and tiny cell size places them, however, with the bacteria, and their place in the world's evolution of life is a whole separate story.

Because red algae lack any motile cells, they also seem more akin to these blue-green bacteria. On the other hand, red algae have complex life cycles; too complex, in fact, to relate here. And red algae vary so much in perceived color that only a few appear truly pink or red to the casual observer. Perhaps it's easier to identify the reds by what they are not, for once you've learned to recognize the browns and greens, the reds are whatever doesn't fit there. There are no other groups of sea-inhabiting algae that are obvious without magnification.

Let's look at a few examples of red algae. One group, the *Porphyras,* are small to medium-sized, flattened, pink to dark red tongues or blades, usually attached to rocks but sometimes also to larger algae. Porphyras are noted for their edibility and are deliberately cultivated in shallow inlets in Japan. Similar

Collecting seaweed in April of 1988, Bun Lucas (Kashaya Pomo/Bodega Miwok) stopped to explain that kelp stem was used as a hookless fishing line in his grandfather's time. A rock with a natural hole served as a sinker when tied about two feet from the end of the stem, to which an abalone bait was also tied. The fisherman rode a driftwood log "boat" from which he held the line, waiting for a cabezone to eat the bait, then pulling the fish in. Kelp stem was also baked and eaten.

—Beverly R. Ortiz

❦

Bun Lucas gathering kelp.

looking reds but with near-black color and a rough texture are called dish-rag seaweeds and belong to the genus *Gigartina*.

Larger, slimy, flattened blades—often so slippery that they cause you to slide off tide pool rocks—belong to *Iridophycus*. *Iridophycus* is so named because the blades reflect different colors that are iridescent from different angles. Beware of these, for their slipperiness can cause serious injury! Another nonred "red" is *Halosaccion* or sea sacks: small, brownish-green, saclike plants in shallow tide pools that are filled with water inside.

Many red algae grow as the finest black, pink, or red lacework, the plant body consisting of numerous delicate branches. Many of these make beautiful specimens mounted on stiff paper, but please don't remove them; use only those that are already washed up and stranded on beaches.

Finally, we have a large group of reds known as coralline algae. These unusual plants secrete a lime skeleton, have jointed branches, and are bright pink when alive but quickly bleach white when removed from water. (Also some corallines grow as pink crusts on rocks or other algae). Corallines are not only vivid splashes of color in tide pools, their lime skeletons are important in the formation of tropical reefs. In fact, many tropical coral reefs and atolls are as much built from numerous coralline algae as they are from the coral animals that also live inside lime skeletons.

If you look at where red algae live within the tidal zone, you'll see that each kind inhabits a specific zone. Those

living in the shallowest (or upper) parts are likely to be the least red; frequently these, such as sea sacs, appear more off-green or brownish-green than red. Those in the intermediate zones are often vivid red or pink. Those living at the lowest levels of the zone or beyond where the tides occur have the deepest colors; frequently these will appear nearly black because of the great concentration of red pigments. Why these color differences? They relate to depth of water, and this in turn relates to the penetration of light. Water filters out light, especially at the blue-violet end of the spectrum. In order to compensate, red algae's red pigments are efficient at absorbing as much blue-violet light as possible (pigments are the color of the light they reflect, not absorb). So with increasing depth, less and less blue-violet light is available and more and more red pigment is present to absorb what little light manages to get through.

All of these examples are merely to whet your curiosity. Next time you visit the beach, look for examples of algae that are washed ashore or better yet, visit a tide pool and note the zonation of its seaweeds. Can you tell which ones are likely to be greens? Which are browns? Which are reds?

*Linda Yamane*

Living near the ocean, I've come to rely on its presence and especially like visiting the beach when the tide is low, walking the rocks and watching for olivella or abalone shells, the wealth of my ancestors. These were traded with

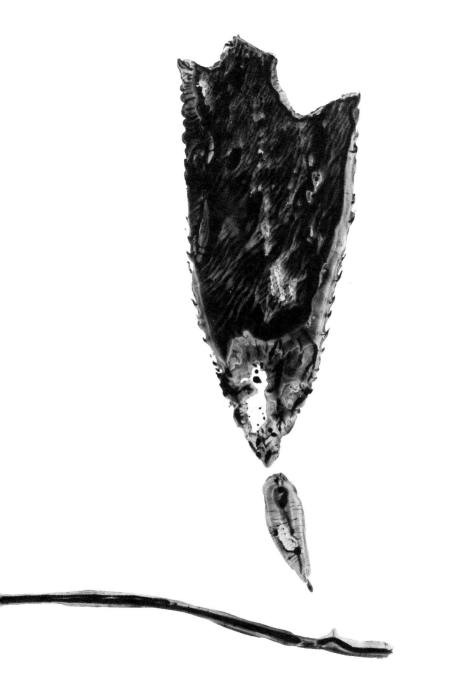

other peoples and meticulously shaped into pendants and necklaces, used to ornament both people and baskets.

I watch, too, for sea lettuce exposed and shimmering green on the rocks. It must have been important in the old times in my area, for one village name translates "the sea lettuce place where people live." I don't gather sea lettuce often, but when I do I like to eat it fresh, especially in soup or salad. The flavor is mild, not salty or "ocean-flavored" as one would expect. I think it's the texture I enjoy most, smooth and with a delicate crunch. In the old times it was used by Rumsien Ohlone people for wrapping around sea foods to keep them fresh; or was dried, then eaten whole or ground into flour. When I've dried sea lettuce, it becomes tough, but rehydrates almost immediately and expands to remarkable size when placed in water. This time around most of its green pigment is gone and so is its crunch—in fact, it's almost gelatinous. I imagine the flour would be a good thickener for soup or stew and wonder if that's how it was used long ago. This I may never know, but knowing that it was important in the past gives me reason to honor and enjoy it today.

Seaweed was once an important trade item, and agreements and arrangements were made so that inland groups could gather it from the beaches at certain times and places. In return for this privilege, coastal people were paid with appropriate gifts.

Though the tradition of gathering and using seaweed was broken in my family, many California Indian people still actively harvest seaweed, continuing traditions established in their childhoods, including the use of special seaweed gathering baskets. For some, seaweed was a staple food when they were growing up, and families spent days at the seashore, gathering enough to last throughout the year.

Before being brought home, the seaweed must be dried in the sun. Later it is cooked in a variety of ways—mostly roasted or fried. In times past, as today, traditions varied from group to group. Seaweeds were cooked on hot ashes or hot coals, pulverized, and mixed with other foods like acorn mush, sometimes steamed. However prepared, seaweed remains a gift well-used and well-received, a food enjoyed in the present that is rich with memories of the past.

# Mosses

*Glenn Keator*

The thick green cushions of mosses are conspicuous in winter, seldom noticed in summer. Among the first plants to colonize the land over 400 million years ago, mosses have changed very little. They occupy the same ecological niches as lichens: rock faces, bark of tree trunks and twigs, barren soils of sand dunes, arctic tundra, alpine situations in high mountains, and new rocky roadcuts and cliff faces. In such places these two groups of organisms grow side by side, often competing for the same light, water, and minerals. Lichens may be the first to colonize if it's a new rock face in full sun; later, as the rock crumbles or cracks, mosses gain a roothold. On shaded forest floors, or on the bark of trees in deep shade, it's often mosses that predominate. Mosses also tend to thrive best in the spray of waterfalls, but don't count lichens out entirely!

Mosses are unique in their life histories and structure. A single microsopic spore starts the process. When it rains, the spore grows a tiny, thin, green branched thread, that looks to all the world like the ancestral green algal strand which first made the transition from water to land. If all goes well, the green thread produces many minute buds that grow into a thick cushion or mat of leafy stems. This forms the moss known to most people. The many kinds of moss are all distinguishable from each other by details of these leafy cushions: sometimes the cushions are dense with truly minute leaves; other times they're more open with leaves resembling miniature pine or fir trees; still other times they carry nearly flattened leaves along creeping stems; yet other times the entire branch system looks like a miniature tree. Leaves are usually paper thin, consisting of one sheet of green cells, but the midrib may be thickened, and the texture may vary from smooth to rough, or the leaves may be thickened at the base or the tip. Leaf shape can be nearly linear to almost circular or anything between, with smooth rounded ends or sharp spinelike tips, and with toothed or smooth edges.

These moss cushions have amazing recuperative powers. Usually we associate mosses with wet, shaded places, but in fact they'll grow anywhere they can get enough light, even on sand dunes or deserts. All it takes is water: after it rains, moss cushions absorb water like a sponge and plump up, turning from dull green, black, or brown, to bright green. Immediately they "come alive" and start photosynthesizing, then shut down again when desiccated. Because of these powers, moss cushions sometimes create beds for young seeds of flowering plants to start in.

Once a moss cushion reaches maturity it fruits. While some mosses make conspicuous green "flowers" at branch tips, it is usually difficult to see the microscopic reproductive structures even with a hand lens. Whether obvious or not, branch ends will carry tiny, globular boxes in which sperms are produced, or minuscule, hollow flasks that hold a single egg each. After a drenching rain, the film of water covering a moss cushion provides the medium for sperms to swim to eggs. This journey may be amazingly difficult, for a sperm may have to travel from the branch of its parent plant to the branch of a neighboring plant, and its only guide is a hormone released by the female.

After the egg is fertilized it grows not into a new moss cushion or even a green thread but an entirely different kind of plant called the sporophyte. To the observer, it doesn't look as though there's a new plant, but merely that the leafy cushion has sprouted a different structure, for the sporophyte plant grows directly from the branch tip where the egg was fertilized. Sporophytes consist of a slender stem—green at first and later turning brown—topped by a bulbous or swollen capsule. The capsule is covered by a cap (the calyptra) that pops off as the capsule turns brown. The change from green to brown signals that the sporophyte is ready to shed its spores and die: each capsule contains thousands of microscopic spores.

But it's not that simple to get the spores on their way; first the calyptra has to pop off, then a little lid at the far end of the capsule (the operculum) has to open. Under this lid lies a complex set of "teeth" (the peristome) that are sensitive to changes in humidity. These work something like a camera lens, opening when the air dries through shrinkage, and closing when the air is humid by absorbing water. Each time the teeth open, air currents shake spores out the end of the capsule, and the process may be repeated dozens of times.

Mosses are such a constant feature of tree bark and shaded banks that they're taken for granted, but when it comes to uses, there are few. Perhaps the most notable mosses are the sphagnums that grow in certain acid bogs. The acid properties of sphagnum make it useful as a disinfectant; dried it becomes the source of peatmoss, a useful soil additive that absorbs and retains water well in porous, loose soils.

Although mosses are poorly understood and difficult to identify, they provide endless fascination of rich detail under a lens and are wonderful fun to observe on wintry, wet days. Then their tufted, green-brown and gray-green

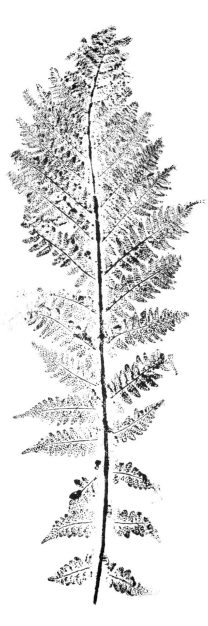

cushions turn to emerald and dark black-green, and their miniature landscapes come alive with activity from furious reproductive growth and the busy antics of animalcules that live within their spongy labyrinths.

# Ferns

*Linda Yamane*

Ferns have an appeal that makes them everyone's friend. They enrich the landscape with a sense of grace and vitality. They also have served a number of practical needs throughout Native California. The two ferns found at the Marin Headlands—bracken and woodwardia—each have a long history of relationship with Indian people.

Let's start with the bracken fern. The large, branched fronds have been used for everything from umbrellas to roofing to the linings of berry baskets, earth ovens, and acorn-leaching pits—and the young, unfurled fiddleheads can be eaten cooked or raw.

The rhizome is also widely used as a pattern material for basketry. Actually, it's not the entire rhizome that is used, but a dark, flat, woody band sandwiched within the mucilaginous interior. It's a messy job to extract and clean it. You have to scrape and scrape and scrape some more, but after being soaked with acorns for blackening, the processed strands make beautiful, black basketry designs.

The woodwardia, or chain fern, also

long basketweaving tradition

Both bracken and woodwardia ferns are indispensable

furnishes a basket material used by weavers in northern California. Within the stem of each mature frond are two long, narrow strands. The stem is pounded, and the two uniform strands are removed, then dried and dyed a reddish brown color with alder bark. Woodwardia does not have the strength to hold up to the rigors of basketry twining on its own, so the dyed material is used as an overlay material, resting atop sturdier conifer roots. Both bracken and woodwardia ferns are indispensable ingredients in a long basketweaving tradition—used, appreciated and cared for by generations of weavers in the past, in the present, and still to come.

The preparation of basketry materials requires specialized knowledge and skills passed down from weaver to weaver for millenia. The processes require patience and commitment, and this intensely personal interaction has resulted in a unique relationship between the plants and their weavers. Weavers care for and are attuned to the needs of the plants, visiting them year after year in various seasons. The plants are not carelessly or wantonly cut or unearthed, but carefully harvested in the proper season, with consideration and in ways that ensure their continued well-being. They are tended, talked to, thanked, and sung to. They are not used against their will, but are partners in a destiny established in the distant past and carried forth today.

*Glenn Keator*

Bracken fern is the world's most consistently successful and widespread species, occurring on all continents save Antarctica.

As befits such an international traveler—it ranges from far north to openings in tropical rainforests—bracken is a rapid colonizer with superior abilities to spread vegetatively. Even though the sori—fertile areas along the edge of the underside of the frond—are relatively rare and so seldom bear spores, it takes only a few fronds in each colony to make sufficient spores to float on winds

to new places. Meanwhile the original parent plant is constantly spreading itself over ever-larger turf by deep-seated underground rhizomes (creeping stems) that branch and rebranch in a rich tapestry. This process of claiming new territory continues until the rhizomes encounter obstacles, such as rock outcroppings, or areas shaded by shrubs or trees where the plant cannot compete for light effectively. Thus you see bracken in open situations: it thrives in grasslands, open chaparral, or cleared areas in woodlands and forests.

A number of key features help identify this fern: each large frond (often standing two to four feet tall and a foot or more across) appears distinct and independent from others, because the fronds are produced at intervals along the horizontal rhizomes hidden beneath the soil.

Fronds die back in late fall, often becoming handsomely burnished before drying up, and are replaced by a whole new set in spring. The young fiddleheads (coiled and curled-up new fronds) identify this as a fern, for all ferns go through the process of unfurling their brand new fronds. When fully developed the bracken frond describes a broad, nearly equilateral triangle.

Few ferns provide food for man, but bracken is an exception albeit a minor one. Several different cultures, including Native Americans and Japanese, seek out the tender new fiddleheads in spring; they're considered a special delicacy as a cooked vegetable. If you try this, be sure to pick only the newest, fully coiled-up fiddleheads; for the older fronds—especially uncooked—are high in carcinogens.

Bracken is also noted for its starchy and fibrous rhizomes. Fibers have been used for strand and twine when other materials are unavailable, although this is seldom the case in California. And the starchy pith that can be beaten out of the rhizomes is noteworthy as an emergency food, particularly in the dead of winter when all of the plant's surplus food is stored there. Few peoples utilize

bracken rhizomes for their basic starch, however, since preparation is difficult and there are usually better substitutes. But for the New Zealand Maoris, bracken served as a staple, since few other native plants provide reliable and abundant starch.

The Maoris are Polynesian, like those who settled such far-flung places as Easter Island and Hawaii. Polynesian wayfarers always carried basic food plants to their new home-lands, including taro (*Colocasia esculenta*) and sweet potatoes (*Ipomoea batatas*). New Zealand's mostly temperate climates, however, made growing taro impossible (they require mild tropical climate) and restricted sweet potatoes to the most protected areas, so bracken provided a much-needed substitute. California, however, never needed to resort to bracken. In our kindlier environ-ment, cattails, pondlily (*Nuphar poly-sepalum*), and various bulbs were plentiful and more easily processed.

# Horsetail

*Linda Yamane*

Spreading prolifically along creek-sides and in other moist places, horse-tails burgeon in their active season, delicate, green filigree meandering at will. I love their primeval presence, the sense of lushness—of life—that they add to the world.

I love them too for their surprising usefulness. The silica-rich stems of the horsetail and scouring rush are wonderfully sturdy and abrasive—and make a great "sandpaper." Not only that, the segmented stems separate conveniently at the joints, resulting in compact pieces that are both easy to handle and to store.

Throughout California these abrasive stems have been used by Indian people, especially for finishing wood work such as arrows. The southern Ohlone also use the black underground stems, flattened and dried, for patterning material in basketry, easily identified by the characteristic vertical ridges also present in the fresh stems. It's a miracle, isn't it, what nature provides—neatly packaged and ready to use.

*Glenn Keator*

Horsetails and their brethren, the scouring rushes, are among our most ancient land plants. Going back almost as far as mosses and liverworts to a time just after plants invaded the land some 400 million years ago, horsetails have changed little in all that time. We know from fossils that horsetails were once joined by many relatives that grew as woody vines or tall trees, but these larger forms have been extinct for a very long time. Evidently the smaller, nonwoody horsetails and scouring rushes, all in the genus *Equisetum* (literally horse hair in Latin), are adapted to such a wide range of conditions and are such strong competitors that they've held on all these years.

35

Horsetails and scouring rushes live in sandy soils along streams and rivers, by freshwater marshes, or by seeps. They often keep company with moisture-loving ferns, such as lady and chain ferns.

They are rather strangely constructed: underground, they consist of many feet of branched and rebranched horizontally-trending stems (rhizomes) along which roots are borne. These rhizomes grow so vigorously in moist, good soil that they'll penetrate every possible corner, thus making themselves a thorough nuisance in gardens. The above-ground plants consist of jointed, hollow, tubular, ribbed, green stems. At each joint there is a sheath formed of blackish to brownish scales (the true leaves that have lost their chlorophyll) and a whorl of feathery, spokelike, green side branches. The latter give the frothy look that so characterizes horsetails (scouring rushes usually lack these whorls of side branches). At close range, each green stem joint has several vertically arranged ridges and furrows; the ridges are impregnated with rough, sandlike silica, giving these plants the unusual ability to act as sandpaper or as scrubbers in scouring pots and pans. Although stems are supposed to provide "emergency food" according to several books, you must peel off the outer silica-rich layer first. You'll then find that this laborious process leaves you with precious little, for most of the stem is hollow.

This tubular quality of the stems gives them great flexibility and allows oxygen to reach the roots in saturated, oxygen-starved soils. Each joint is separated from the next by a fragile disc of tissue; it's here that the plumbing system (vascular tissue) forms complex patterns so that side lines can head into the radiating side branches while other strands continue their journey upwards into each ridge above. Because the ridges of one joint alternate with the furrows of the joint immediately above and below, the vascular tissue must grow in a complex pattern to accommodate such architecture.

Horsetails evolved long before seeds or flowers were invented; like the ferns, they still rely on spores for reproduction, and also like the ferns, they have an alternation of generations. Spores are produced by the thousands in compact brown to black cones. In the scouring rushes, these cones appear sporadically at the tips of green stems, but in our horsetail, they're only produced in earliest spring, and then on separate stems.

Horsetail plants change dramatically with the seasons. In late fall, cold temperatures cause the collapse and disappearance of green stems, but the underground rhizomes carry on all winter long. Then, just as days are warming and lengthening, rhizomes send up many short, whitish stems tipped by dark brown cones. Soon after the cones release their spores, their stalks wither, but at the same time, new green stems are emerging. These green stems are what we usually notice, and are present through spring, summer, and early fall. What is confusing to the novice is that cone-bearing stems and green stems emerge from different places on the soil's surface. (The rhizomes wander widely.) It appears as though we have two entirely different kinds of plant: one with chlorophyl-less pale stems and conspicuous cones, the other with feathery, bright green stems.

Each cone is a marvel of design. The outside is covered with dense, spirally-arranged, polygonal scales. These scales bear several long, saclike sporangia (spore containers) on their inner side and are attached to the core of the cone by a slender stalk. Pull off one of these scales, turn it upside down, and observe the spore sacs under a hand lens. Spores are girdled with a pair of elastic ribbons (elaters) that expand when moistened and contract when dried. These ribbons help flick the spores out of their casings when the air dries temporarily between rains.

Despite the use of spores for reproduction and the ephemeral alternate generation that grows from them, horsetails have succeeded well for 400 million years. Although we talk of the benefits of reproduction by seeds, there is an obvious lesson here: horsetails have managed well for far longer than any seed plants have been in existence.

# Monterey Cypress & Monterey Pine

*Glenn Keator*

Historically, few trees grew naturally at Marin Headlands, nor did they for that matter in what is now San Francisco. The original environments in both places featured frequent heavy winds, particularly in summer, and these winds beat down tall plant growth. Winds operate in two striking ways to oppose tall plants (the taller the plant, the stronger the wind; winds are minimal close to the ground): they dry delicate growing tips of shoots by carrying moisture away, and they blast these shoot tips by abrading them with sand and other particles.

When outsiders first decided to settle here they were inclined to change the environment to suit their needs, and that meant taming the winds or reducing their impact. Golden Gate Park was once open sand dunes; the park was made possible by planting trees and shrubs resistant to stiff winds. In other areas, too, wind breaks were widely established by planting such trees. Where did these trees come from? From other windy environments in California, where trees had already adapted to strong winds. The Monterey Peninsula turns out to have such an environment, for great forests of Monterey pines (*Pinus radiata*) and Monterey cypress (*Cupressus macrocarpa*) naturally grow there, often on precipitous bluffs overlooking the ocean where salt spray and strong winds are real deterrents to plant growth.

These two trees assumed great importance in creating suitable habitats for plantings in Golden Gate Park; indeed today we still see the skyline dominated by these trees. The same is true for buildings protected from winds at the Headlands. These trees also have the virtue of growing rapidly to full size; in fact, it's for just this reason that Monterey pines have found widespread use throughout Bay Area landscapes as, for example, in the Oakland hills.

Fast as these trees grow, they are destined to short lives (as measured by other trees, that is); life expectancies seldom exceed 100 years. Even before reaching this age, these trees begin to show unpleasant symptoms of senility. In particular, the larger limbs grow brittle and are prone to breaking off violently during winter storms or especially strong winds. People are now learning that neither Monterey pine nor cypress is appropriate around houses, close to heavily occupied garden spaces, or even as a permanent wind break. One of the hardest problems with which San Francisco's Recreation and Park Department is grappling is the replanting of wind break throughout Golden Gate Park. Many trees there are overdue to die.

Monterey cypress and pine also have a splendid and curious distinction: both are relatively restricted in their natural homes, yet they've become the most widely planted timber trees in New Zealand and many parts of Australia: sort of a cultural exchange for the Australian blue gum.

Yet for many millenia Monterey cypresses and pines have had a difficult time in competing with other trees in the wild, as California's climate has grown ever more arid. Originally covering large tracts in coastal forests, today Monterey cypress naturally occurs at just two sites; the area around the famed Seventeen-mile Drive in Carmel and at Point Lobos State Reserve just south of the Carmel River. Monterey pines have fared only a little better, occurring in large forests around Año Nuevo (northern Santa Cruz/southern San Mateo counties), the

entire Monterey Peninsula, and the southern Monterey coast around Cambria. Both trees are limited to shallow, rocky, or sandy soils close to the coast, where low nutrients and heavy winds have eliminated any would-be competitors. Both also depend heavily on summer fogs to keep them growing; inland the hot, dry summer conditions would soon kill them. Fog is used very effectively in these pine or cypress forests, for the droplets condense easily on needles and drip to the ground, adding precious moisture otherwise unavailable.

Monterey cypress is the biggest of our native cypresses, often reaching 20 to 30 feet tall and with an almost equal spread. The heavy branches stand out straight from the trunks, and from a distance give a tiered or pagodalike appearance. Winds further shape and prune the branches into truly picturesque shapes, as graceful and evocative as though a master sculptor had shaped them.

Monterey cypresses are further distinguished by the shallow strips of brownish bark (often weathering gray), the bright green but tiny, scalelike leaves (use a strong hand lens), and the circles of brown, globe-shaped seed cones. The tiny, scalelike leaves reduce the amount of surface area exposed to strong winds so that as little water as possible is lost during even the most blustery days.

The seed cones—large for cypresses, hence the scientific name "macrocarpa" for large cone—are covered with tough shield-shaped scales, each ending in a short protuberance (the umbo). Scales protect on their inner side several small,

winged seeds. These cones are perpetual features of the tree, for they seldom open on their own and remain tightly attached to their branches. Try to remove one, you'll find they're almost impossible to pull off without cutting or twisting.

Monterey pine is a moderate sized tree with a crown shape that reflects local wind conditions. The branches are thickly clothed with dark green needles borne in clusters of three, giving the trees a healthy and restful look. The most distinctive feature is the seed cones, which are borne in whorls (the species name "radiata" alludes to this) and pressed against the branches they grow on. Each cone, from three to five inches long, displays a spiralled arrangement of woody, flattened scales, each scale ending smoothly or in a short, spinelike protuberance.

These seed cones stay tightly strung to the parent tree—try twisting one off to experience the strength of this connection—and seldom open unless there have been unusually hot daytime temperatures. Only the heat of fire or the death of the tree by other means allows the seed cone scales to dry and shrivel, thereby opening the cones and dispersing the seeds within.

So we see a common thread that has adapted both Monterey cypress and pine to their natural environments: reforestation after fire by cones that otherwise remain steadfastly closed. This is why you will see that natural stands of Monterey cypresses or pines appear uniform in height and age, for all of the new saplings grow at about the same time during the wet spring following a summer or autumn wildfire.

# Blue Gum or Eucalyptus

*Glenn Keator*

Visitors to California and California residents alike often think of blue gum eucalyptus as indigenous. Nothing could be further from the truth, for the blue gum hails from southeastern Australia and the island state of Tasmania. There it grows in open forests, sometimes

with other eucalypts, on rocky granitic slopes, or on fertile embayments of land between fingers of bay and ocean. Contrary to popular belief, which holds that the soil beneath eucalyptus is sterile, blue gum forests in Australia have their own companion plants thriving beneath their shade, with several attractive shrubs such as hopbush and ferns such as bracken. In California, gardeners find

it next to impossible to plant beneath blue gum, and although the strongly scented leaves are blamed (and may be part of the reason), the principal culprit is the deeply delving roots, which are greedy for water. It is true that the camphor-like oils in blue gum leaves are poisonous to other plant life and may inhibit their growth. But, the reality is that abundant watering or naturally occurring rains leach away these oils, removing them to deeper places or carrying them away to streams.

Blue gums were first introduced to the state in the late 1800s, with the idea of creating quick-growing, profitable wood lots for timber. The saplings responded to their new homeland with enthusiasm— after all, California's Mediterranean climate with hot, dry summers and cool, wet winters is somewhat similar to the climate of their native country—and the new woodlots sprouted hundreds of tall trees in a few years. The problem came later when the wood was ready to be harvested, for then it was learned that blue gum wood becomes hard—unworkably hard—after it has lain around for any time. Today, the wood is still an excellent source for fires, as long as it is immediately chopped into usably-sized pieces, but other uses are impractical. Australia is home to several hundred more kinds of eucalyptus, but today nobody is seriously trying them. This seems almost strange, for in their homeland, several species of eucalypt are in fact in great demand for their beautiful wood, particularly the jarrah from southwestern Australia, where the climate most closely matches our own.

Meanwhile an even greater irony has occurred in the exchange of "economic" plants with Australia, for the Aussies have taken to heart our own native Monterey pine and have planted it far and wide in woodlots and managed forests.

Despite their tendency to "take over" an environment, blue gums are fascinating trees with many fine attributes. While young, they're easily identified by their pairs of broadly elliptical leaves covered with a bluish powder—the reason for the common name. These juvenile leaves are broad in order to absorb more light under the canopy of other already established trees. As they mature, the higher branches switch over to making single, vertically oriented, sickle-shaped leaves, a ploy to reduce water loss by presenting less surface directly to the sun. The camphor-like substances in the leaves have had wide use as pharmaceuticals in the preparation of cold and cough remedies. These same substances are used by the tree to carry on chemical warfare against would-be browsers, and may also slow down competing trees in the close neighborhood.

Flowers of the blue gum announce its membership in the myrtle family, Myrtaceae, a family prominent in tropical rainforests as well as in the droughtier habitats across much of Australia. Blue gum flowers open by a cap popping off the top; this cap represents the sepals of the flower and is the reason for the generic name "eucalyptus" which arises from two Greek words: eu—true or well—and calyptus—covering or cap. When fully open, each flower consists of a

shallow cup filled with nectar attractive to bees and surrounded by myriad long white stamens, which lend the flower its color (petals are missing). After pollination, the stamens fall away, leaving the nut-like fruit to develop. When ripe, these woody capsules fall to the ground or sway about in winds, and as they dry out, a second cap which covers the capsule pops off, so that seeds may be shaken out.

Blue gums are also noted for their long, peely strips of grayish-brown bark. As these long strips peel, at times crashing to the ground in wind storms, they reveal a new white bark beneath. It's by the combination of bark pattern and color, flower arrangement and color, and shape of seed pods that most eucalypts are identified. Californian gardeners grow several other kinds of eucalyptus, but these seldom present a problem, or go "wild" as do the blue gums.

# Willows

*Linda Yamane*

Where there is water, willows are sure to be plentiful—and they have been a blessing for Indian peoples who have used them in so many different ways. Willow bark has long been esteemed as a treatment for colds and fever; and it's no wonder, for it contains the same active ingredient as modern-day aspirin. The inner bark was also used to make rope in some areas.

Because of their flexibility, willow poles were a popular construction material for houses and other community buildings in the past—and in fact are still used for traditional structures today. Narrow branches were well-suited for arrows, game pieces, and any number of other useful things.

The slender rods of certain willow species were prized and are still harvested for basket making. The willows used in basketry have attractive bark and narrow pith. If the pith is too large, as in most willow species, it is easily exposed when scraping the stick to uniform diameter, and will break more easily.

Willow withes are harvested in the spring if the bark is to be removed, for in this season the sap is running and the bark slips easily. Winter cuttings will keep their bark, and some baskets, such as fish and eel traps, are only made from the winter willow.

Willow sticks are used in two different ways in basketry—either whole or split lengthwise. When used whole, they are sometimes scraped to counteract the natural taper and achieve a uniform diameter from end to end. Some baskets don't require that the sticks be scraped and they are used as is. Whole sticks are used as the foundation rods in coiled basketry or as the frame in twined ware. Sticks are split lengthwise, either in halves or thirds, then split again or scraped to remove the inner wood. This leaves a very thin and flexible strip that can be used as the sewing material in coiled baskets or for twining. The roots are used in the same way, being split and trimmed and dried for many months before use.

> "The willows, they're water plants. I don't mean they just need water, all the plants need that. I mean, they grow where the water is—a water place; around the springs and on the creeks. If the willow's there, the water's there even if you don't see it and it looks dry on top. It's there, I know it.
>
> "Because it's just about everywhere and it grows lots, it means that it's for everybody to use and there are all different ways that people can use it. That's why in the early days we did so many things with it, and that's why it's such a powerful medicine, and why you can fix so many things with it."
>
> —Laura Somersal, Pomo/Alexander Valley Wappo basketmaker

❧

Elsie Allen with cut willow sticks.

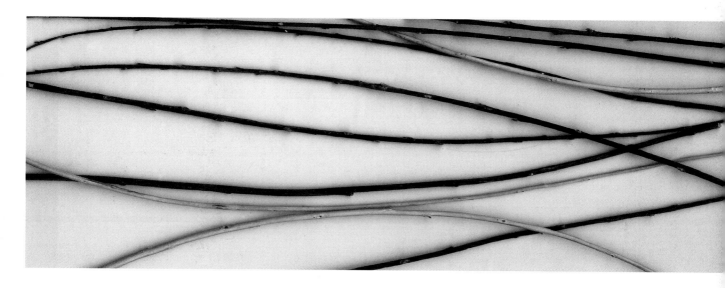

*Glenn Keator*

Willows are nearly universal across the northern hemisphere, and are indicators of permanently high water tables. Growing along the edge of marshes, streams, and rivers, they dominate over other vegetation, or when mixed with other trees form an understory to taller kinds, such as cottonwoods, maples, and alders.

Willows have long been regarded for their special properties: as trees or giant shrubs, they grow rapidly and easily from cuttings and suckers as well as seed, and they stabilize slopes or create cover for birds in a way second to none. So easily do bundles of willow stems strike root, that a watery extract from the bark is useful in inducing other cuttings to do the same. In addition to rapid growth, willow stems are noted for their pliability; bend them any which way and they'll accommodate. This quality has made them first rate for creating baskets, a quick framework for small huts, living fences, and all manner of other implements and furnishings. Willow twigs are also the classic "witching sticks" used by diviners of hidden water; twigs are said to bend toward the earth when a witcher passes over a place with a reliable, high water table. Finally, willow bark is the source of salicylic acid, the active ingredient in such modern pain killers as aspirin! The word for this acid comes to us from the Latin word for willow, *salix*, and in another form this Latin root gives us the name of the Marin town of Sausalito (Spanish *sauce*, willow tree; *-al*, grove of; *-ito*, little).

Willows are interesting to observe at just about any time of year. In winter, the bare twigs are recognized from afar by their yellow, orange, or reddish bark (according to kind), and the long, pointed buds which are pressed tightly against stems. In spring, the first stirrings of longer, warmer days herald the unfurling of now enlarged fuzzy buds—the stage called "pussy" willow for the soft, furry covering. Male willow trees open their buds to reveal upright catkins of yellow flowers—each flower a group of yellow-tipped stamens brimming with pollen shed to spring breezes. Female willows open their buds to upright catkins of greenish flowers—each flower a green ovary tipped by a short style and feathery stigma to catch pollen flying by. Since wind is used to move the pollen, willow flowers have lost their petals. The arrangement of male and female flowers on different trees assures cross-pollination.

branches and leaves at a fast pace thanks to their reliable and constant water supply. As a result, willow leaves are thin and seldom protected from losing water, although the undersides (where tiny pores in the leaves open during the day to absorb needed carbon dioxide) may be pale or covered with wooly hairs to reflect heat and help retain moisture. Finally, as fall days are ushered in and night temperatures plummet, willows lose their leaves, sometimes quickly during blustery storms, other times gradually during gentle rains. Coastal willows seldom turn color, but if you travel to the high Sierra in October, there you'll see the subtle yellows and tans of willow leaves before they're shed.

# Coyote Bush

*Glenn Keator*

Coyote bush is a common evergreen shrub, often associated with coastal chaparral that occurs along rocky bluffs and promontories. Its growth pattern varies from low, creeping, semiprostrate plants to fully upright, densely-branched shrubs that reach up to eight or ten feet tall. Form follows conditions, for the low, prostrate kinds grow on exposed bluffs, where winds keep them low (they also are partially dwarfed genetically), while tall kinds grow inland. The tallest coyote bushes are, in fact, probably those in the process of being shaded out by invading trees, and the extra tall branches are an

Just after willow flowers have fin-ished—their movement of pollen unim-peded by leaves—the new leaves are released from their separate buds. By now the seed pods on female trees are burgeoning, and soon they split open to release hundreds of tiny seeds, each borne by a wad of cottony, white hairs. This fluff is widely carried, for willow trees depend upon long distance dispersal to get their seeds to new poten-tial sites, and suitable habitats are often miles apart and separated by high ridges. This same fluff renders female willow trees a nuisance in the landscape, for the fluffy seeds easily penetrate crevices and openings and intrude upon country households.

Willows wear their leaves the rest of spring and summer, making new attempt to reach for as much light as possible.

Coyote bush is densely clothed in round-elliptical glossy leaves. Leaf margins bear occasional teeth, and leaves become resinous and fragrant on hot days. The fragrance is reminiscent of sweet incense. Volatile oils evaporating from leaf surfaces are responsible for this quality, but also make coyote bush a liability because it becomes so flammable. Such oils are responsible for a suite of features in many California native shrubs that favor survival in a droughty climate. Perhaps foremost, the oils are distasteful to the browsers and munchers of the animal world. Then, there's the phenomenon of leaf cooling, for on hot days when oils are evaporating most rapidly, this form of "plant sweating" cools the leaves just as our own sweat cools our skin. Finally, many volatile oils are harmful and inhibitory to seeds of competing plants, so that when they have penetrated the soil, they prevent competition for the precious commodity of water.

Coyote bush blooms in fall, September to November, with separate male and female plants bearing separate male or female flowers. Without flowers, the sex is impossible to tell. Belonging to the daisy or composite family, coyote bush clusters its myriad, tiny flowers into tightly packed heads, scattered lavishly over the plant. Unlike many daisies, however, there are no showy outer ray flowers that function as petals to attract pollinators; only tiny, star-shaped disc flowers. Male flowers are cream color, with protruding pale yellow stamens;

female flowers are whitish, owing to the long white hairs that extend well beyond the flowers proper. These same white hairs later efficiently transport fruits for long distances on winds.

Because the seeds are carried so well for considerable distances, coyote bush is one of our "pioneer" shrubs, moving rapidly into prairies that are no longer grazed, for example. Or they are frequent sights in vacant lots that have not been tended for some time. Coyote bush begins a process botanists call "succession." Where many a cleared field is left to its own devices, changes in the vegetation take place over several years. The first stage is generally the reappearance of woody shrubs, with coyote bush leading the brigade. Later, if local conditions are appropriate, shrubs are overshadowed by trees, but in many places such as the Marin Headlands, this probably will never happen, for the strong winds there prevent much tree growth.

# Blue Bush Lupine

*Glenn Keator*

Blue bush lupine is prominent in that community of plants dominated by shrubs and clinging to rocky coastal bluffs: what botanists call coastal scrub. Here it joins company with coyote bush, coffee berry, poison oak, and bush monkeyflower. Blue bush lupine stands out when its multiple branches carry candle-like clusters of blue-purple flowers in May

and June. These flowers open over a period of several weeks, decorating the hillsides when many early spring wildflowers have already passed.

Each flower is—on close inspection— a miniature version of the garden sweet pea, thereby disclosing lupine's relationship to that far-flung plant family. The upper petal—banner—stands erect and acts as a flag to potential pollinators, advertising that nectar is available. The two side petals that protrude forward are called wings, in allusion to their shape. Pull these wings off and you'll find a canoe-shaped structure between them: the keel. The keel actually consists of two petals that are joined or fused most of their length. Pull down on this keel and the reproductive parts—stamens and pistil—are revealed. Lupine flowers are usually serviced by bees, particularly the heavier kinds such as bumblebees. The weight of such insects as they land in search of nectar causes the keel to be lowered, while the stamens and pistil remain fixed in place. This results in the bee's belly being dusted with pollen—if stamens are shedding pollen just then— or being cleaned of its pollen from another flower—if the pistil's stigma is receptive. To add to the lure of bringing bees in to look for food, lupine flowers release a sweet, subtle perfume.

Like other members of the pea family, lupines bear tiny knobs or nodules on their roots. Inside each nodule lives a colony of microscopic bacteria that perform a miraculous service for the roots: they convert nitrogen into a form that plant roots can use in their search for nutrition. Nitrogen is the single component most often in short supply in soils because it is needed in such large amounts for healthy growth. Nitrogen is the first substance listed on a bag of commercial fertilizer, for the same reason. While these bacteria confer a special advantage on lupines, at the same time they benefit from the safe home they have inside the root and the supply of sugars sent their way as byproducts of lupine leaves photosynthesizing. When lupines die, the extra nitrogen is eventually released into the soil for use by other plants, so lupines improve their local environment by enriching soils.

Curiously, the name "lupine" comes from the Latin word for wolf. When lupines were first named, it was believed that because they are often found on barren soils they had robbed the soil of nutrients in the fashion of marauding wolves in a farmyard. Of course, we now know that the opposite situation is true: lupines grow in nutrient-poor soils because of their capabilities of obtaining nitrogen from their bacterial partners.

# Sticky or Bush Monkeyflower

*Glenn Keator*

Receding into its background in late summer and fall when leaves begin to shrivel, sticky monkeyflower is a conspicuous element in coastal "soft" chaparral in winter and spring. Then it sends out a fresh batch of shiny green, lance-shaped leaves and by spring's end, a riot of showy, orange, two-lipped flowers. The leaves grow sticky on hot days because they exude a varnish-like substance making it difficult for insects and browsers to chew on them and sealing in precious water that would otherwise evaporate. Even though these methods of preventing water loss are effective, in long and unusually hot summers, leaves shrivel and drop one by one in order to ensure survival until fall's first rains.

Sticky monkeyflower has especially attractive and interesting flowers that appear by late spring. Each orange flower consists of a long, flared tube ending in two "lips:" the upper two petals creating an upper lip and the lower three petals a lower lip which also serves as a landing platform for bees. Orange is an unusual color in our flora, being common otherwise only in poppies and paintbrushes; yet although there are few pollinators that favor orange, the color the human eye sees is undoubtedly complex, consisting of a yellow base with orange pigment laid on top. The yellow is clear and attractive to bees, while perhaps the orange serves to attract butterflies.

Pollination by bees is probably the most useful way of getting things done, since the reproductive structures—stamens and pistil—inside each flower are clearly designed for them. The lower lip is often decorated with darker orange spots as a guideline to the nectar which is hidden deep within the tube. At the entrance to the tube, two white stigma lobes protrude; these are the first structures a bee is likely to hit as it probes for nectar. In doing so, it may be carrying a load of pollen from another flower and the pollen is combed off the bee's back by those stigma lobes. At the same time, a remarkable thing happens: those lobes are sensitive to touch and quickly close when stimulated by contact. You can demonstrate this easily by touching the middle of the lobes with your finger or a pencil; within a couple of seconds the lobes have closed! Now as the bee probes deeper into the flower tube, he picks up a new load of pollen from the two pairs of stamens hidden inside. When he backs out to visit a new flower, there is no way the pollen from the present flower can be accidentally spilled onto the stigma since the stigma is now tightly shut.

All of this intricacy seems a bit overdone until the idea of cross-pollination is explained. Throughout the living world inbreeding (or "selfing" in the case of flowers) leads to genetic weakness. It is for this reason that humans have strong codes against incest. When the code is broken, such as among some of Europe's royal families where intermarriage is common, certain genetic weaknesses—such as hemophilia—have appeared frequently. We also see the results of inbreeding in lines of purebred dogs where genetic weaknesses are commonplace. So in light of our understanding, it's not difficult to imagine why flowering plants have established their own means of preventing genetic weakness by promoting cross-pollination between flowers of different plants and discouraging self-pollination.

Another compelling reason for favoring cross-pollination is the variation seen

"*Plants are thought to be alive, their juice is their blood, and they grow. The same is true of trees. All things die, therefore all things have life. Because all things have life, gifts have to be given to all things.*"

*—William Ralganal Benson, Pomo*

in the offspring. Wide variation, as for example in flower color, plant size, and leaf shape, is common in many of our most successful species. Such variation also allows individuals to adapt to almost imperceptible differences in habitat and environment.

Monkeyflowers are named for their alleged resemblance to monkey faces; scrutinize the pattern of the petals as seen in a face-on view next time you find the flower and see if you can recognize a face.

*Linda Yamane*

Monkey Flower
    stands
    tall
    in the sun
with orange-colored
    blossoms
    full
    of nectar
    sweet
    and waiting
    for a tasting.
    Yum!
And sticky leaves
    make great
        fun.
But Monkey Flower
    has a serious
        side, too.
Raw leaves
    and stems
        when crushed
        and poulticed
        will defend
        a wound and
help it heal.

And if that's not
        enough
from our sticky friend,
        there's even more
        it will do.
You can eat it!
        Just the young
        leaves and
        stems—either
        cooked
        or raw,
        whichever
        you please.
(But don't forget
        to say
        "Thank you!")

# Poison Oak

*Linda Yamane*

Like most hikers, I learned early on to be mindful of poison oak. I've never had a reaction to it and suspect that, like my father, I am not allergic. But I always figured, "Why take a chance?" and have learned to be aware and to avoid contact whenever possible. I've seen it in all its seasons and varied locations from shade to sun, witness to its variability—brilliant red-ness, lush green-ness, bare stick-ness, large-leafed and small. It amazes me that poison oak can be so varied. On dry, sunny, chaparral slopes the leaves are thick and compact, their smallness a remarkable contrast to the large, soft-bodied leaves found when poison oak grows in moist, shaded places.

For most people, poison oak is a nuisance, for others a serious threat to their health and well-being. Like mosquitoes and ticks, it's one of those things whose purpose on earth we find easy to question. Have you noticed that we tend to measure things in terms of how they benefit or threaten us? If something benefits or pleases it is "good," if there is potential for harm it is "bad."

Poison oak has become a lesson in perspective—cause for reflection—a reminder that we do not always have to be the center, the standard for whom all things are measured. For over the years I've seen deer grazing in poison oak thickets. For them it is not an enemy, but a sustaining friend. Likewise, deer have sustained California Indian people for uncounted generations. So I have learned to embrace poison oak, if only figuratively, appreciating it as the friend of a friend and accepting that there is value in all things, even when I cannot see it.

In fact, older Indian friends have told how, in their childhoods, mother or grandmother would have them eat or chew a leaf or more of poison oak to prevent colds, or for immunity. One of these friends, no longer living, once told me that she, however, would never consider feeding her own grandchildren poison oak leaves, for fear of the possible ill effects to her young loved ones.

And thus, over the years, traditions of use have changed, it seems; for although poison oak leaves are said to have once been used to wrap around acorn bread—in my own area as well as elsewhere in California—I don't know of anyone doing so today. Likewise, the use of poison oak's slender shoots in basketry, or its black juice for dyeing basket materials or for tattoos, doesn't seem to be practiced any longer. Yet knowing these things gives cause for seeing poison oak—and the people who have so creatively used it—anew.

*Glenn Keator*

Popularly dreaded and reviled, poison oak is at once a plant to know well, steer clear of, and learn about. Not an oak at all, poison oak belongs to the cashew or sumac family, Anacardiaceae; the individual leaflets of the three-part leaves resemble certain deciduous oak leaves, especially those of valley and Garry oaks. These same leaves are the best warning sign from early spring to fall. Each leaf is composed of three separate leaflets, each leaflet obovate, but coarsely scalloped and/or lobed. The size of these leaflets changes with the season (small when they first emerge) and quite dramatically according to environment: exposed to full sun they're one to two inches long, but in deep shade, they may exceed five inches. They're bronzy-red when they first emerge, but soon turn a glossy green, remaining that way to summer's end, finally turning a brilliant scarlet before falling to the ground. In years of severe drought, leaves fall by mid-summer, a way of keeping plants from losing excess moisture when every drop counts.

The leaves and twigs of the poison oak plant carry an oil called urushiol, a severe skin irritant to many. Even when the twigs are bare in winter, they're dangerous to brush against, although urushiol is produced in largest amounts

## Toxicodendron diversilobum

A FEW SENSIBLE REMARKS          301

"Why take a chance?"

in spring as new growth resurges. Carry a bar of mild soap when you walk, and you'll be able to rinse the oils off your skin (be sure to do this within 15 or 20 minutes of contact). Don't forget that the oils may remain on clothing or shoes unless they're washed, and that pets' fur may carry the oils from an encounter in the brush. Pets seldom break out, for their fur coats protect their skin from direct contact.

The most severe damage occurs when poison oak is burned and the smoke inhaled. And even those who claim complete immunity may be susceptible later in life; immunities have a way of changing from good to bad or vice versa, so it pays to learn the multiple personalities of this plant. Once a rash has broken out—from two to three days after contact—it cannot be spread by scratching; the reason it appears to spread is that different parts of the skin may break out at different times.

Poison oak can be found in just about every plant community where shrubs or trees occur, and even on exposed coastal bluffs where winds buffet the branches causing them to lie prone to the ground. In chaparral or coastal scrub, poison oak is normally a dense shrub growing to eight or ten feet tall. In woodlands and forests, it is not uncommon to see it turn into a woody vine which climbs tree trunks in search of more light. Redwood trees may wear a girdle of poison oak soaring to 40 or more feet in its pursuit of sun.

Besides the characteristic leaves, poison oak sends out short, nodding racemes of whitish-green flowers in mid-spring; you'll actually smell their sweet perfume before you notice them, but don't fear, for this is the one part without urushiol. In fact, the flowers drip with nectar, and their sweet scent is irresistible to bees; since the plant needs their services, it cannot poison its helpers. Flowers are succeeded in summer by small, whitish berries eaten by birds. Once again, the berries are edible—at least to birds—and the birds render the service of softening seeds as they pass through the gut and moving them to new places where they're eventually excreted. Even in winter, poison oak is recognizable if you carefully study the branch and twig pattern. The smooth, pale brown bark and many long, upright branches contrast with the several stubby side branches or twigs.

Poison oak is indeed a close relative of poison ivy from eastern states, and is also related to the poison sumac from those areas. All contain the same oil, urushiol. Plants seldom arm themselves for no reason; it's likely that urushiol serves as a deterrent to would-be browsers, for when the oils are ingested they produce a severe and immediate irritation impossible to ignore. Other more distantly related members of this family seem to have similar caustic oils. Cashew nuts have an outer covering which may blister the skin, although the base of the fruit is fleshy and edible (cashew apples are widely eaten in tropical countries). Similarly the skin of the mango, and even the flesh for those particularly sensitive, may irritate the mouth and its delicate linings.

Despite its evil reputation, poison oak is very much a part of California's foothill environments, from the seashore to the innermost Coast Ranges and from the exposed headlands in full wind to the deep shade of redwood forests. Perhaps we should learn to keep truce with it, for it provides us with beautiful seasonal change, fragrant spring blossoms, and fruits that serve as food for birds.

# Berries

*Linda Yamane*

People love sweets, and the California landscape doesn't let us down even when it comes time for a treat. Wild blackberries, huckleberries, gooseberries, elderberries, and a host of others are still widely gathered and enjoyed today.

In the past, berries were favored throughout California and different tribes developed distinctive styles of baskets for gathering them. They are usually smallish in size to prevent the fruit from being crushed under too much weight, and designed to be worn around the neck, leaving the hands free for picking.

Of course the fruits were eaten fresh in season, but blackberries are also known to have been dried and traded by people living along rivers where blackberries are abundant. And abundant they were— women sometimes filled burden baskets with upwards of a hundred pounds, picked in a single afternoon! Blackberries were also a favorite food of grizzlies, who were often encountered amid massive blackberry thickets, making it wise to take a dog along to scare the bears away.

The Yokuts developed an elaborate and ingenious method of drying and pulverizing blackberries, then mixing the meal with water and drying it again in small cakes for storage and future cooking into blackberry jam. The secondary drying of the blackberry paste prevented the fermentation that would occur with simple drying and storage.

Wild strawberries, small in stature but ever generous in sweetness and flavor, were enjoyed, anticipated, and honored with special springtime dances and festivities. The tradition continues today with annual strawberry festivals held in central California, enduring acts of thanks and celebration for what the earth provides.

*Glenn Keator*

Although called woodland strawberry, in coastal areas this plant is often found on rocky banks and along the edge of dense shrubberies rather than under woods. It is able to survive in these places because of the dense fogs that help cut the full brunt of intense summer sun.

This delightful ground cover resembles garden strawberries in most particulars: each leaf is divided into three broadly rounded leaflets, each with a distinctive and attractive feathery vein pattern; the leaves are nestled into neat rosettes; long runners (or stolons) extend from these rosettes to establish new plantlets; and the white blossoms resemble miniature single white roses. Stolons are a clever vegetative ploy to spread more plants locally, assuring that once a pioneering individual has established itself it will quickly be able to make many copies of itself and co-opt as much of the local environment as possible. And, as plants spread themselves about, each new plantlet quickly anchors itself to the soil by roots, and the whole colony is further stabilized.

The fruits—tasty and sweet—look like miniatures of the garden kind. In fact, it is from these tiny "berries" with their concentrated flavor that other forms of this very strawberry have been selected for gourmet tastes. These selected forms are often called by their French name, "fraises des bois," meaning "strawberries of the woods." These other forms come from Europe and were once held to belong to a different species, but recent studies show the close relationship between the plants that occur naturally in California's lowland woods and those from eastern North America and Europe. The reason for the high value of these flavorful fruits is that they pack a lot of "punch" into a small space: larger fruits, including most garden strawberries, substitute a lot of water for concentration of flavor, with predictable results.

The berries of woodland strawberry are actually not what they seem, for botanists have a very precise definition of a berry: a single fleshy fruit with many seeds inside. By botanists' definition, bananas, tomatoes, and eggplants are actually berries. One other essential part of the definition is that the ovary (the female part of the flower that surrounds and protects seeds) becomes the fleshy layer in which the seeds are embedded or encased. In actual fact, the

things that pass for individual seeds in strawberries (the tiny brown flecks) are really single-seeded ovaries. You can see the origin of these tiny ovaries if you look at a blossom: there, the center of the flower is filled with numerous tiny green pistils, each of which later turns into a brown, one-seeded ovary. This means that the fleshy wrapping on a strawberry is not the ovary but something entirely different; it wasn't even there at the flowering stage. What happens is that the supporting tissue at the base of the flower—what's called the receptacle—gradually enlarges and engulfs the tiny one-seeded ovaries. By the time strawberries have ripened, the receptacle has completed its development and now is a sweet, red inducement to partake in the pleasures of this "accessory" fruit.

Whether a plant packages its seeds in a fleshy ovary or receptacle is immaterial to the way the seeds are dispersed. All fleshy-fruited plants are designed to attract hungry animals, be they birds, mammals, or both. The color, odor, and taste of fleshy fruits is aimed at whatever group best serves the fruit's purpose of being eaten. In any case, the tiny seeds resist being digested and are later excreted, while the fleshy parts have provided a meal. The whole reason for having such edible fruits is simply to move the seeds to a new place, in order to grow away from the parent plant. In the process, the digestive juices have also softened the usually hard seed coat, and this promotes germination.

Wild blackberries are familiar to anyone who goes berry picking in summer and fall. They are among our most

prolific and delicious berries, abundant and juicy in years of plentiful spring rains, more inclined toward seediness and smaller fruits in droughty years. Wild blackberries are common sights along most coastal backroads and fields cleared from forests or along forest edges. Sometimes they create impenetrable hedges because of their exuberant abundance. In fact, it is this latter feature that makes most blackberries attractive to gather from the "wild" rather than grow at home.

Actually we have two distinctive species of blackberries: the most prolific kind is Himalaya berry (*Rubus discolor*), with large leaves divided fanwise into five leaflets and stout thorns along its stems and leaves; the other kind is our native dewberry (*Rubus ursinus*), with leaves divided into threes and slender prickles along stems and leaves. The Himalaya berry, as its name implies, is not native, but was likely introduced for its delicious berries. Because our coastal climate fits its needs, it has "escaped" from cultivation in many places and continues to invade new territory. Once established, it's almost impossible to eradicate, for the branches arch over and plant themselves as the tips touch ground; the roots are strong and vigorous with the ability to send up new shoots in place of the old; and the birds eat and disperse the seeds as they feast on the fruits. The native dewberry is also prolific in its own right, with many of the same attributes as the Himalayan kind, but a bit less aggressive and overpowering.

Blackberries first bloom at the end of spring, with a profusion of showy white

56

to pale pink flowers. Each flower is designed like a small single rose indicating family kinship. Bees are particularly fond of the nectar and so pollination is usually carried out with expediency. The several separate tiny pistils of the flower each develop into a miniature fruit of their own called a drupelet: each druplet has a center stone-covered seed (these are what get in between teeth after eating a bowlful of berries) surrounded by sweet, acidic, dark purple pulp. Because there are many pistils in each flower, the many drupelets are clustered together to make what we call a berry. Botanically, as mentioned previously, berries are defined as fruits with a single ovary containing many seeds, and so the blackberries are properly called aggregate fruits, referring to the clustering together of several separate (albeit tiny) pistils.

Blackberries are closely related to raspberries and thimbleberry (*Rubus parviflorus*), but differ in the way the ripe fruit breaks away when picked. With blackberries, the center "core" to which the drupelets are attached comes along with the berry; but in raspberries the fruits pull cleanly away, leaving a thimble- to cone-shaped core behind. Thimbleberry is a native shrub in coastal California, with softly fuzzy, maple-shaped leaves, and has dark red fruits of excellent flavor. Its fruits behave like those of the raspberries. Salmonberry (*Rubus spectabilis*) is yet another common coastal California plant that belongs to the blackberry category: it is a shrubby plant with weak spines and arching branches, and its salmon egg-colored drupelets pull off with the core attached.

# Stinging Nettle

*Glenn Keator*  🌿

Newcomers to California's streamsides and riparian corridors don't readily forget their first encounter with stinging nettles. Stems are armed with copious stinging hairs, each a potential weapon against would-be browsers. Each hair is like a miniature hypodermic needle, bursting on contact, and issuing contents that blister the skin: a substance similar to the formic acid of ants is involved.

The painful welts that result gradually subside, but the results may linger for several days. Strangely, a frequent companion plant—red elderberry (*Sambucus callicarpa*)—may relieve these stings; crush a leaf and apply to the affected area and the sting temporarily disappears. Dock—a common weed in similar habitats—is also known to relieve such stings.

Stinging nettle occurs in dense "hedges," particularly on sandy banks along shaded streams, flourishing and competing with such plants as scouring rush (*Equisetum hyemale*) and California wild rose (*Rosa californica*). In one season its stalks may soar to over 6 feet, carrying the ovate, coarsely-toothed leaves to the very tip. Nettle makes drooping clusters of inconspicuous, greenish, wind-pollinated flowers.

Nettles are renowned in Europe for their use as strong bast fibers (in the stems) for creating twine and rope (similar to those found in hemp) and for nutritious, vitamin-rich leaves that make a good-tasting cooked vegetable. To use nettles as food, wear heavy gloves and pick only the newest tips of the shoots in spring; cooking destroys the stinging properties.

Stinging nettles belong to a family named for this stinging property: the Urticaceae. The same name is used for caterpillars and other insects armed with stinging or irritating hairs, commonly found in tropical rainforests.

*Linda Yamane*  🌿

Stinging nettles! With respect and deference we wisely leave space between nettles and ourselves wherever we meet, for a close encounter of the tactile kind is not soon forgotten. But not all friends are "touchable"; and though less appealing in the conventional sense, nettles nonetheless have much to offer and their most valuable gift is well-hidden. I love to see them standing tall and green, adding lushness to their already wet surroundings. In their youth and maturity they have a puzzling combination of strength and delicateness, their leaves suggesting frailty yet always seeming to defy gravity by remaining staunchly parallel to the ground.

I often take special note where nettles grow and vow to come back in the winter, after a frost, when the sap has dropped and the nettles are no longer "armed and dangerous." They are not so handsome then—in fact, they're ugly. They're dry and dull-looking, standing grey and shriveled and dead. But inside

"TAAWAX"

each wasted stem is a miracle waiting to happen. String! Generations of central California Indians have valued nettles (as well as milkweed and dogbane) for the sturdy long fibers that run the length of the stem, sandwiched between bark and pith.

After cutting the stems, we scrape the bark, then crush the stem to create a lengthwise split. I usually split the stem into two halves. Starting at one end, we then break the pith at one- or two-inch intervals and peel it away. After working the fibers a little to soften and separate them some, they're ready for twisting into cordage. I love twisting the fibers against my leg, working my hands and watching the strands ply one on the other, transforming miraculously. And I think about the beautiful nets and bags our Native ancestors made from handmade string. I know the patience, skill, and effort required, and I feel rich with pride and admiration.

# Soaproot

*Linda Yamane*

It is always a wonder to me that a plant as unglamorous as the soaproot can have so much to offer. Often there is very little of the plant showing above ground, and you will need a sharp eye to spot the brown fibers peeking out at the surface. When the leaves emerge in the spring, their greenness calls attention to them as they elongate, wavy and slender, tempting passing deer who often nibble the tapered leaves down to blunt ends. Eventually the leaves dry, shrinking and losing body as they drop lower to the ground. At this time you can look for the tall central flower stalk that continues to stand even after its usefulness has passed. Looking down closely at the base of the plant, you may even have to push aside a bit of soil to reveal the fibers that will verify your encounter with soaproot.

Soaproot has so many uses it's hard to know where to begin, but by far the most common use today is for making brushes—sturdy, brown-fibered brushes for sifting acorn flour. Soaproot brushes range from stout to slim, depending on the style of the maker, but regardless of style the brushes are made in the same way and are jewels of ingenuity.

The fibers are first removed from around the bulb, care being taken to keep the fiber bundle intact. Then they are arranged in the desired width and thickness—the tapered end of the bundle will be the handle, the curved end will do the brushing. Next comes the big job of removing the dirt and debris from the fibers. Running an awl repeatedly through the fiber bundle while holding the bundle firmly together and shaking the fibers seems to work best. During this process the shorter fibers are also removed, so they won't end up in our acorn flour later! When the bundle is thoroughly cleaned and the fibers neatly aligned, they are tied firmly together along the portion that will serve as the handle. Most people penetrate the bundle during the tying, figure-eight-style, so that the inner fibers will be secured and not pull out with later use.

To make the handle, we boil the soaproot bulbs in a little water until the flesh is soft. The bulb is constructed of layers, similar to an onion, but the fleshy layers also contain fibers that must be removed. The cooked bulbs, when cooled, are rubbed against a sieve or open-work basket, which allows the soaproot pulp to pass through to the opposite side, while trapping the fibers on the working surface, thus freeing the pulp from the fibers. The pulp is then applied to the brush handle, using a little water to smooth the surface, and allowed to dry. This step is repeated until the handle has reached the desired thickness.

Soaproot brushes are used to clean out mortars and baskets and especially for brushing acorn flour from the sifting tray. They are indispensable—both esthetically and culturally—and a symbol of California Indian culinary art.

Now the soaproot didn't get its name for nothing, and anyone who has tried using it for washing will have learned it is not second-rate! The crushed bulb, when mixed with a little water, makes a sudsy lather that is also a fine shampoo.

The list of uses goes on and on. Large quantities of crushed bulbs were used in the past to "poison" fish in dammed streams. The cooked bulb can be eaten, though it is not a favorite of mine, and the pulp provides a serviceable, though water soluble, glue. The leaves were even used to wrap around acorn dough before it was baked into bread. Soaproot may be modest in appearance, but it has got to be one of the most versatile plants I know.

The path winds on,
　　we find ourselves
　　　　among familiar friends.
Coyote Bush looks the most comfortable
　　on this dry September day.

Say—I'd sure expect to see Soaproot
　　here too
　　and I think fondly
　　　　of its sturdy brown fibers
　　hiding so cleverly underground
　　　　holding within
　　　　　　a bulb—
unglamorous, it's true—
but a bulb that yields
　　　　a most marvelous glue!
And I think of the graceful
　　soaproot brushes I've made—
each fitting so easily in the hand—
and I keep looking for my friend.
　　　　Soaproot,
I know you must be here
　　I can feel you should be
　　but I just can't find you...
oh—there you are...now I can see!
It's hard to find you
　　with your leaves brown, too.
Enjoy your rest
　　for the days will come
　　when winter rains will beckon
　　and you must push forth again
　　your wavy green leaves
　　which I will see so easily.

Native Bunch Grasses live here, too.
　　They're new
　　　　to me
and yet I see them waving
　　　　cordially
　　as we walk by.
I hear their dry, scratchy rattle

and feel good to know
that there is still
　　　　someplace
where they can grow.

*Glenn Keator*

Soap plant, soaproot, or amole—it goes by all these common names—is perhaps our most abundant bulb. The onion-sized bulbs store ample food and water for vigorous growth, following the long, dry summer and fall. Rosettes of leaves typically emerge after the first few rains. These leaves, even without any flowers, are easy to differentiate from all others, for they're tinted bluish green and have an undulating or wavy edge. Like so many other lily relatives, leaves are long, narrow, and strap-shaped, with clearly parallel veins.

Leaves don't show much sign of changing once they've come up, and even by spring, little has happened. But some time in mid-spring, a narrow, asparagus-like stalk appears in the middle of the leaf rosette. Just like asparagus, this stalk has green scales and many tiny buds that only slowly expand. By late May or June, the stalk has completed its extension, and now branches widely and openly. Inland, these flowering stalks extend up to three or four feet high, but near the coast they remain relatively low, seldom surpassing eighteen inches.

Many people never see soap plant's blossoms, for they open late in the day, one at a time on each side branch of the flowering stem. In fact the species name "pomeridianum," means "post meridian," in allusion to the late afternoon opening of flowers. The white flowers remain open in the evening, when they appear suspended in mid-air, for they are attached to very slender stalks that disappear against the evening's black background. These flowers have evolved, of course, to display well in the evening light. Rather than depend upon bees, which are active only during the day, soap plant blossoms attract small night-flying moths, flies, and beetles. Since there are few other night bloomers, soap plant has the pick of the available pollinators for that time slot and so avoids the problem of competing with all the other day-blossoming flowers.

Each soap plant blossom consists of six nearly identical narrow white petals, each striped on the outside with a central purple line. The six stamens point forward, as does the stigma at the end of the pistil, right in the path of a pollinator seeking nectar from the petals' bases. Later, a near globe-shaped seed pod swells, turning from green (in its

"When I was a small girl, I went on root digging trips with my mother and helped her to collect plenty of roots to dry for winter use. These would be gathered in baskets. Some were cooked whole, or sometimes we pounded them up and cooked them like mush...."

—Marie Potts, Maidu

unripe condition) to dark brown, as seeds ripen inside. The seed pod splits into three chambers, each with several near-black seeds inside, which are spilled by winds moving the now brittle flowering branches to and fro.

Soap plant's leaves gradually turn brown and shrivel, and by summer's end the flowering stalk topples over; but before this process is completed, most of the food and water have been stored in the bulb. Once the leaves and flowering stem are no longer obvious, the only indication of these plants in the field is the long, brown fibers that surround the bulbs: these sometimes extend up through the soil to the surface, particularly in places where the bulb is not deeply buried.

Soap plant's bulb is instructive of how bulbs are put together. The coarse, brown, outer "husk" of fibers represents the skeleton of the vascular system in highly modified leaf bases. Inside this lie a series of several overlapping, thick, white, fleshy scales, where water and food are stored. These scales occur in the same pattern seen in onion bulbs and represent additional modified leaf bases. Fibers and scales are attached at their base to a circular corky disc which represents a very compressed, scaled-down stem. It is from this stem that a whole series of roots grows after the ground has been saturated by fall's first rains. In the very middle of the bulb, a conspicuous bud sits protected by all of those scales. It is this bud that holds the potential for next year's growth. Contained within it is a series of new leaves and within them a future flowering stem in miniature.

Soap plant and its myriad sister bulbs thus initiate next year's leaves and flowers at the end of this year's season. Should this year fail of its promise for abundant moisture and warm temperatures for optimal growth, next year's blossoms will reflect the fact. Thus, bulbous plants are at once tied to the conditions of the prevailing season but are also influenced by the mix of last year's growing conditions.

# Manroot

*Glenn Keator*

Manroot is aptly named, for it stores its food and water in an enormous root the size of a person. Of course it takes several years before the root grows this large, but a mature plant can be almost impossible to dig out due to its massive root. The manroot goes dormant all winter. The winter rest is important in this mostly tropical family that also contains gourds, pumpkins, squashes, melons, and cucumbers (it is also called "Indian cucumber"). Susceptible to cold, it survives California's winters by going dormant.

That enormous food store coupled with numerous dormant buds means that when the weather first warms in earliest spring, the buds are primed to go. Each plant sends up myriad young shoots that look a bit weird, for they're mostly bright green stem, with a few tendrils from the young leaves that wave around. If manroot shoots find no suitable support to cling to, they soon flop over and run along the ground, but normally they grow by shrubs or trees, and those curled tendrils are able to find and twine around these in a short time.

Growth is very rapid, and soon the shoots that looked so gangly have grown into several feet of vine, each tendril a perfect green spiral designed to help the vine climb. Climbing brings the leaves into closer proximity with full sunlight, and all the while the stems are growing like gangbusters, the leaves have been unfolding at a furious rate as well.

Soon—usually before mid-spring—flowering stalks appear, one per leaf. Each short stalk may carry ten to thirty cream colored or white starlike flowers. It's great to watch these open, for most flowers are male only, with fused stamens in their center but no trace of a pistil. Because female flowers have a pistil with an inferior ovary—the ovary shows as a bump behind the other flower parts just as with a zucchini squash—it's easy to tell them apart.

Sometimes the entire flowering stalk may carry no female flowers—particularly at the first part of the season—but soon one or two female flowers open at the base of most flower clusters.

Again, watch what happens as the petals fade, for now a truly remarkable transformation begins. The original green ovary is covered with what looks like a three-dimensional tapestry of green hooks, the whole thing about the size of a pea. Soon however the ovary grows and grows, going from pea size to the size of a rounded cucumber, all covered with green spines.

Whether these spines are needed to dissuade animals from eating the developing seed pod is dubious, for the pods themselves taste bitter and contain rather poisonous seeds inside. If you were to cut open a near-mature pod, you'd find the interior lined with a thick, whitish pith with a few rather large, bean-shaped seeds embedded in it. But now the real fun comes, for as the pods develop a strong water pressure in their covering, the weakest part of the pod— the tip—suddenly gives way in a sort of vegetable explosion. It's wise not to stand in front of the pod when this happens, for the seeds are ejected with considerable force and owing to their large size can actually hurt.

Finally as the drama of seed pods exploding subsides, the manroot goes into a rather quiescent summer period. If the soils now become dry, so do the vines, turning from deep green to golden yellow-tan, but in the fog belt, the cool summers usually keep vines healthy until fall. Then at last, as fall days turn temporarily hot, the final process of going dormant takes hold, and the whole vine—often twenty feet or more long—withers away while the food and water is stored once again safely below the soil in the root.

Manroot announces its membership in the gourd family (to which the cultivated cucumber also belongs) by the palmately lobed leaves, the viny stems with curled tendrils, and the unisexual flowers, the female flower with an inferi-

or ovary (as with squashes). This important family has some of our very best vegetables and fruits, including watermelon (from northeast Africa), other melons (from the Mideast), cucumber (from India), squashes (from Meso-America), and pumpkin (also from Meso-America). Gourds belong here as well, but they're used for the hard shell around the fruit pulp rather than for food.

Despite the edibility of many "cucurbits," our native manroots—despite the name Indian cucumber—are a decided exception; their seeds are highly poisonous, and the fruit pulp and casing is also inedible. The Indians did use the plant, but not for food: instead the large, beanlike seeds were crushed and tossed into streams to stupefy fish, so they could be easily scooped up. Evidently the poisons in the seeds did not affect the edibility of the fish flesh. Other plants used in similar fashion included bulbs of soap plant and seeds of California buckeye (*Aesculus californica*).

# Wildflowers

*Linda Yamane*

Wildflowers! They surprise and delight us with explosions of color and form that appeal to our esthetic senses and are capable of filling us with wonder like children again. California Native peoples have long enjoyed the wildflowers, some celebrating with springtime flower

dances, ready to receive the many gifts they provide.

Many California wildflowers produce edible seeds that were important staple foods to people in the past. Describing their encounters with coastal Native peoples, the early diaries of Europeans repeatedly include accounts of wonderful-tasting seed foods. Baskets filled with seed meal, porridges, loaves, or small cakes were presented alongside many other gifts offered the foreign visitors.

Some seeds were parched by tossing with hot coals or sand in a shallow basket tray. These toasted seeds were then ground into flour. The seed meal could then be eaten without further processing, or made into various kinds of seed cakes. Some seeds are so oily that they can be shaped, especially when warm, into cakes that will keep their form without any further ado. Some seed meals were baked into cakes in earthen ovens, or cooked as porridges. The cooking of porridges was in specially-woven cooking baskets that would hold water. Since baskets cannot be placed on a fire like pots and pans, Native peoples devised a way to put the heat into the basket, in the form of hot cooking stones. These round, igneous rocks are heated in a fire, then removed and rinsed of ashes and other debris before being placed in the cooking basket containing the seed meal and water. The stone is then stirred until the heat transfers to the contents, at which point another stone is added and stirred, and another and another, until the porridge boils and thickens.

This is the way that acorn was cooked in the past and still is today—for many an Indian person will tell you that it just doesn't have the same flavor when cooked on a stove.

# Douglas Iris

*Glenn Keator*

Douglas iris is our showiest native iris and easy to grow in the garden. Look for its floral displays as early as the new year during mild winters, for then the first buds unfurl in warm pockets along the immediate coast. Douglas iris grows from a series of shallow rhizomes that grow outward from the center as radiating spokes. As the colonies grow ever larger the center dies out, leaving an encircling wreath of leaves and flowers. Each year the circle grows larger. Very old colonies eventually break apart into separate clones, but by then they've lived several years.

Douglas iris is an enduring plant, for it passes through summer droughts and winter chill without dying back; it keeps handsome, glossy, sword-shaped leaves all year. These overlap in a fanlike fashion at the base—something botanists call equitant—and are a mark of other members of this farflung family. The flowers are carried on a foot-long stalk, at first protected inside a pair of conspicuous floral bracts. Two to three buds sit inside, each taking its own turn at flowering. The flowers are exquisite at close range:

many say they're reminiscent of orchid blossoms. Near the coast they are a deep, rich, blue-purple, while inland they tend to pale lilac. Every once in a while, an individual appears with pure white flowers, but it's the same as the blue-flowered plants in all other respects.

Iris flowers are elaborately put together: the three "falls" (sepals) look like turned-down petals; these alternate with the "standards" (true petals), that stand stiffly erect. Above each sepal is a long slender stamen, and above that is what looks like a third petal. This innermost whorl of floral parts is the style branches masquerading as extra petals; in fact, each style branch overarches and hides (and protects) the stamen below it. Look carefully underneath the style branch just below its split tip and you'll find a minute, triangular flap of tissue. This is the only part of the entire style branch receptive to pollen; in other words, the stigma!

Now let's look at how these structures function. The sepals' tips are gentle landing platforms for bees, and are provided with stripes and lines leading inward to the tube of nectar at the base of the flower. As the bee scrambles down, his back first encounters the stigma flap. If he's just come from another iris with a load of pollen, this flap brushes it off his back. As he probes still further down, his back now encounters a stamen loaded with pollen, and a brand new mass of pollen is combed onto his back. Finally he reaches the nectar with his tongue, then backs out of the flower. Because the stigma flap can only move in one direction, he does not accidentally deposit this new

load of pollen on the same flower.

Douglas iris also has hidden attributes. Those shiny leaves are lined with a fine, translucent fiber that Native Americans removed carefully to create twine of high quality. Those same leaves are poisonous as food, and grazers as well as browsers quickly learn to leave them alone. It's perhaps for this reason that some of our coast's most overgrazed pastures still sport healthy colonies of Douglas iris, which—thanks to elimination of the competition through grazing—thrive even better than before.

Because of the great beauty of its flowers, Douglas iris has long had garden admirers. The English were the first to cultivate it, but now at last Californians have discovered it for themselves. It is fairly well established in the nursery trade—at least at specialty nurseries—and should be tried by anyone with an interest in wildflowers.

Douglas iris has also been hybridized with other native irises to produce a truly stunning array of colors: pure white, bronze, ivory, golden yellow, red-purple, all shades of blue and purple, and more. In these hybrids it is truly seen that the original word "iris", from the Greek for rainbow, is well applied.

Two miniature iris relatives should also be noted: blue-eyed grass (*Sisyrinchium bellum*), with delightful umbels of wide-open, saucer-shaped blue flowers centered yellow; and yellow-eyed grass (*S. californicum*), with similar umbels of pale golden flowers. Blue-eyed grasses dot coastal grasslands in April; yellow-eyed grass brightens seeps and springs on rocky promontories in May and June.

*Linda Yamane*

Wild Iris
 with elegance
  blooms
white and lavender
  surprises
But more than
 mere beauty
  is present here
for certain species
 bear
  within
each slender leaf
 white fibers—
semi-translucent
 silky
  thin
Among these
 two
are remarkably
 strong
and by Northern California
  men
 were rolled
  and plied
  on bare leg
 into long
  perfect
   string
What painstaking
  transformation
from delicate filament
to deer snare ropes
fishing nets
 enormous
 and small
rabbit nets
 to stretch
 a hundred
 yards or more

net bags for
 gathering
  tobacco leaves
Signs of wealth
 all of these
  and
 exquisite reminders
that there is
so much more
  than meets
the modern eye.

# Blue Dicks

*Glenn Keator*

Blue dicks is but one of a whole series of corm-forming plants called brodiaeas. The original name honored a Scotsman named Brodie and has become firmly attached to several related kinds of native bulblike plants in the lily family.

Of these, blue dicks is always the earliest to bloom, sometimes opening its first flowers in mid-March. Depending on locale, the floral display may carry on through April and at higher elevations as late as June. Blue dicks is also our most abundant brodiaea, occurring by the thousands after fire but common even under ordinary circumstances. Near the coast it is succeeded later by a miniature form of Ithuriel's spear brodiaea (*Triteleia laxa*) with pale, sky-blue, agapanthuslike flowers. Still later comes the miniature ground brodiaea (*Brodiaea terrestris*), which looks like so many blue stars sprinkled among the drying grasses.

Blue dicks has sinuous, naked flowering stalks to a foot or more (according to soil depth and severity of wind); these carry tight umbels of blue to blue-purple, bell-shaped flowers surrounded by deep blue floral bracts. Each floral "head" may contain upwards of twenty blossoms, assuring a long display. On warm days, these flowers play host to a variety of would-be pollinators from honey bees to hover flies to tiny beetles or swallowtail butterflies.

Finally as the last flower fades, the process of seed production begins; by the time grasses have dried, the seed pods are ready to spill their seeds and the whole plant dies back to take a rest. This summer rest is typical of a whole array of native "bulbs," better termed geophytes (literally, earth plants). Just when the growing gets difficult because of summer drought, the food and water reserves can be hidden several inches underground, where soils remain cool and slightly moist. In this environment the plant prepares for next year's flowers, and by the time fall rains arrive, it's all ready to sprout again and start the whole process over.

Geophytes hedge their bets in reproduction by doing things two ways: every year—if pollination has been successful—a new batch of seeds is produced, as with other flowering plants. But also each year, the bulb (or corm) makes several baby bulblets. This vegetative means of reproduction succeeds in increasing the local population exactly without variation. When Native Americans dug bulbs for food—blue dicks was a favorite—their digging sticks would dislodge the baby bulbs, allowing new opportunities for

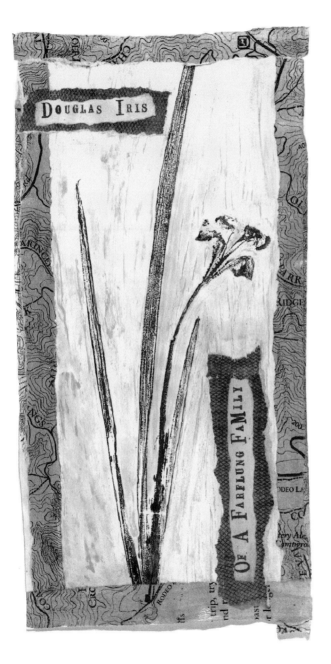

DOUGLAS IRIS

OF A FARFLUNG FAMILY

these to replenish what was taken.

What about the fate of seeds? Well, every year sees a new batch germinate and grow, but at first it's hard to detect the new seedlings, for they closely resemble grass seedlings. And the fact that so few people are acquainted with seedlings of bulb-producing plants, besides their grasslike appearance, is that the seedlings require several years to reach blooming size and so are not a practical means of propagation for those wanting expedient results. Gardeners often fail to grasp that bulbous plants can reproduce from seeds as well as bulbs. Nature makes such bountiful supplies of bulb seeds, though, that patience pays off with an abundance of new plants.

Another reason that bulbs are seldom propagated by seeds is that particular varieties will not "come true" from seeds; that is, the offspring show variation in such traits as flower color, plant size, and number of flowers per plant. For example, rare white-flowered individuals seldom produce seedlings with white flowers, for white is a color that is recessive in genetic terms.

So the story of seeds versus bulbs shows that nature often has more than one way to reproduce herself. New bulbs permit immediate copying of the parent plant for optimal reproduction in a favorable locale; seeds allow for variation in offspring with the chance that many seeds will end up some distance from the parent. When this happens, the conditions in the new locale may or may not be similar to those where the parent grew. Variable offspring hedge bets that some gene combinations will be favorable should the conditions be different.

*"We had many relatives and...we all had to live together; so we'd better learn how to get along with each other. She [my mother] said it wasn't too hard to do. It was just like taking care of your younger brother or sister. You got to know them, find out what they like and what made them cry, so you'd know what to do. If you took good care of them you didn't have to work as hard. Sounds like it's not true, but it is. When that baby gets to be a man or woman they're going to help you out.*

*"You know, I thought she was talking about us Indians and how we are supposed to get along. I found out later by my older sister that Mother wasn't just talking about Indians, but the plants, animals, birds— everything on this earth. They are our relatives and we better know how to act around them or they'll get after us."*

—Lucy Smith, Dry Creek Pomo

❧

*Frank Howe (Modoc) fishing with a net made of native materials.*

*Linda Yamane*

Many of the wildflowers we enjoy today have also furnished important underground sources of foods for Indian people. In fact, Native women spent so much time harvesting edible roots, bulbs, corms, and tubers, as well as the roots and rhizomes used in basketry, that the invading Americans called Indian people "Diggers."

Americans had come west in greater and greater numbers, taking more and more from Native peoples who were finding it increasingly difficult—sometimes impossible—to provide for themselves and their families. Indian people no longer had access to sufficient food resources or safe places to live without fear of molestation. Adults were often killed, their children taken and sold as slaves. In order to justify this cruel treatment, the newcomers developed a dehumanizing stereotype designed to characterize Indians as dirty, uncivilized creatures who did little more than scratch through the dirt like animals for roots and bugs. We may still cringe at the memory, but we also remember that digging was a fundamental and respectful part of Indian life. Some basketweavers today, not ashamed of digging roots from the ground, are beginning to turn it around and say, "I'm proud to be a digger."

California Indian women gathered their cornucopia of bulbs, corms, roots, and tubers with special digging sticks. These sticks were generally two to four feet long, made of hardwood steamed straight, with a sharp fire-hardened point. Some were fixed with doughnut-shaped stone weights.

Because bulbs grow in beds, they were fairly easy to harvest, especially when the ground was tilled in the process, year after year. Today, where ground is untilled and compacted, the job is a formidable one. Sometimes whole hillsides or meadows were filled with these "Indian potatoes." Brodiaea bulbs can be eaten raw, but are sweeter when baked. The bulbs were roasted in large, earthen ovens, built by lining a pit with rocks and heating with a fire, then raking the coals, placing the tubers and bulbs between layers of protective foliage, and covering everything over with dirt. A fire was then kept burning on top for several hours, after which the covering was removed, the sumptuous contents ready for eating.

Wild onions, sweet potatoes, garlic and carrots, camas and brodiaeas, and members of the lily and parsley family were roasted, steamed, boiled, or dried. If dried, they could be pounded in mortars and then pressed into cakes for use in winter. Some people pounded the fresh uncooked bulbs, then mixed the pulp with berries and baked it as small flat cakes. However these foods were prepared, the earth provided a bountiful feast.

# Poison Hemlock

*Glenn Keator*

Poison hemlock, lining roadsides and open fields from late spring through summer, is a common sight. So characteristic is it of California that few realize this tall biennial comes to us from Europe, where it has had long use as a poison. In fact this is the very same kind of plant Socrates is said to have taken when he died. When the word "poison" is mentioned, everyone thinks of plants dangerous to the touch, as with poison oak. But most of our poisonous plants are in fact perfectly safe to handle so long as they're not eaten. Such is the case here.

Another common roadside follower, often confused with poison hemlock, is a relative, Queen Anne's lace (*Daucus carota*). The two show their close affinities by traits that unite the parsley family, Apiaceae. They both have numerous flowers borne in open compound umbels (umbrellalike clusters of flowers, where the flower stalks radiate out from the end of a common stem like spokes of a wheel). Other features common to the family are highly scented leaves and stems, and sheathing leaves often much divided into many separate segments. Queen Anne's lace and poison hemlock both have lacy, fernlike leaves, and both also have white flowers. But there the resemblance ends. Perhaps the best feature for quickly distinguishing the two is the irregular purple spots and splotches on poison hemlock's stems; Queen Anne's lace has no spots, but there are fuzzy hairs on stems and leaves. Finally, the broad umbrellas of flowers in Queen Anne's lace change from flat-topped to basket-like; as the flowers age the central ones are sunken below the level of the outer ones. (Also the odors differ: Queen Anne's lace has a carrotlike fragrance, and poison hemlock has a disagreeable, bitter odor.)

It turns out that Queen Anne's lace is no more than the cultivated carrot gone

"wild." When cultivated plants escape from the garden, they often revert to their original wild form over a few generations. In the case of Queen Anne's lace, the root is woody, not tender and succulent as with the cultivated carrot.

On the other hand, poison hemlock belongs to its own small group, and the species name "maculatum" alludes to the purple spots (immaculate is spotless in English; if we had a word "maculate" it would mean spotted).

Another relative in this group—water hemlock—is native to California and found in slow-moving streams and freshwater marshes. There its tall, flowering stems stand several feet high, but the leaves are more coarsely divided than those of poison hemlock. Water hemlock contains one of the most virulent of all naturally occurring toxins: a piece of root the size of a pea can cause violent convulsions and death. Fortunately, the chambered roots help distinguish this plant, but great care is needed if you're picking watercress, since the two often grow in close association

The whole issue of poisons in plants is a fascinating subject: often tiny amounts of the poison have therapeutic and medicinal effects. The parsley family is famed for its medicinal and toxic plants, but is surprising in that it also contains several edible vegetables and culinary herbs. Consequently, great care is needed in distinguishing which plants are safe to consume and which are to be strictly avoided. Examples of vegetables here include carrot, celery, parsnip, and finocchio. Typical culinary herbs are parsley, anise, fennel, caraway and dill.

# Cow Parsnip

*Glenn Keator*        🌿

Cow parsnip is another common example of the parsley family and it too is abundant in coastal California. At first glance this plant seems very much like poison hemlock and Queen Anne's lace, but look again, and you'll begin to see clearcut differences. For instance, the large, foot-long leaves are coarsely divided for cow parsnip, and begin to look ragged and worn as spring turns to summer. Smell these leaves—odor is very reliable and unchanging for each member of the parsley family—and you'll note a second important difference. Some call cow parsnip's odor rank and unpleasant, others rate it as appealing; but it's difficult to name the odor exactly since our language is poor at conveying the sense of smell. In fact, most of our words relating to smell are comparative; that is, the odor is said to resemble something whose smell is familiar. For cow parsnip there is no direct comparison with anything else. It smells like cow parsnip, and that's all there is to it!

Cow parsnip is a native perennial that takes a winter rest. By early spring the new shoots are pushing up; soon they grow at a frantic pace, so that by April or May the flowering stalks stand six to eight feet. As each leaf is unfurled, more stem grows up, and finally the uppermost leaves—whose bases bulge—are holding the flower buds. The umbrellas of flowers are astonishing, for while poison hemlock flower clusters may measure six inches or so across, cow parsnip makes umbrellas well over a foot in diameter. Look closely and you'll see how pretty the individual flowers are. In fact if you look carefully, you'll notice that the outer flowers—around the periphery of each umbellet—are larger and showier than the inner flowers. This is a case of using flowers for different purposes: the outer flowers are for show (to attract beetles and bees), the inner flowers provide nectar and produce seeds. Such a division of labor is found in several unrelated groups of plants, but reaches its culmination in the design of the daisy or composite flower head.

By late spring the broad umbels of white flowers are everywhere, lighting up rocky coastal promontories and open places in coastal chaparral. In summer the flowers turn to fruits, and the fruits ripen to pretty, flattened discs surrounded each by a narrow white to brown margin or wing. When winds finally carry off the fruits, these wings serve as aerodynamically sound lifters, allowing wide dispersal on air currents. Finally by summer's end the leaves and flower stalks wither and fall to the ground, their energies spent for one more year; but the roots carry on over winter in preparation for next year's floral display.

Cow parsnip has its adherents as a wild "food." Many books comment on the parsniplike roots, which peeled supposedly can be cooked as a vegetable. Whether many of the authors of these books have tried them is doubtful, for the strong smell of the leaves is also typical of the roots. Perhaps someone out there would find them palatable. Another use of note was the drying and burning of the leaves by Native Americans. The ash was considered a good salt substitute, though it's doubtful this use was important along the coast where the real thing was easy to obtain. Instead, it's probable that this use of cow parsnip was important to groups dwelling near mountain meadows, where cow parsnip also grows in profusion.

# Mustards & Radish Weed

*Glenn Keator*

The far-flung mustard family is both commonly found and easy to recognize. Some mustard relatives belong here naturally; others are widespread aliens that have been naturalized for over a hundred years. For example, field mustard was introduced as a cover crop for orchards and vineyards and also as a source of mustard oil, so widely used in culinary preparations. Radish weed is simply the garden vegetable gone wild that has reverted to a vigorous weedy form with woody roots.

All members of the family share a floral blueprint with four separate sepals, four individual petals arranged in crosslike fashion, and six stamens: four long and two short. The cross design invoked the first name used for the family, Cruciferae, or "bearer of crosses." Even the seed pods with their two to four rows of seeds bear an indelible family trait: a papery parchment separates each pod into two identical halves. This parchment remains on the plant long after the seeds' and pods' covers have been shed.

Mustards are annual plants. Starting as low rosettes of leaves that are deeply slashed and roughly textured, the plant quickly sends up one main stem, lined with smaller leaves. Soon the main stem branches several times, and then flower buds appear. Buds open to bright yellow flowers. Flowers are followed by long, slender seed pods that are capped by a blunt, enlarged tip.

By contrast, radish weed, although starting life in much the same fashion, has flowers that may be pale purple, pinkish, bronze, pale yellow, or white. Seed pods that follow are about three times as long as wide, fat where seeds sit inside and pinched in between seeds to give a distinctive appearance. Seed pods break open across the narrow places rather than lengthwise as in mustard.

Radish weed also grows quickly once it has started—provided that moisture is available as it usually is in early spring. At about the time you'd be digging up the leaf rosette to harvest the radish roots, radish weed begins to grow in earnest, quickly sending up a flowering stalk to over a foot. If you've ever left your garden radishes too long, the same thing happens; in fact, it's for this reason that radishes all tend to be harvested at one time, for left to their own devices, they soon "bolt." At this stage, the main difference between the garden vegetable and the weed is the root, for if you pull up a typical radish weed you'll not find a crunchy, fat edible root; instead, there's a tough, woody taproot that is indigestible.

The bright yellow fields of mustard are one of the earliest evidences of spring; in years with mild winters, mustard may blossom in earnest by late winter, signaling the many events to come. Mustard fields are often intermixed with or succeeded by masses of radish weed blossoms, creating fetching displays of spring flowers. It's easy to forget that neither radish weed nor mustard is native.

Because mustards and radish weeds are annual, their seeds are geared to make the most of a decent wet period, and seedlings germinate nearly any time there's abundant moisture and mild temperatures. Slow at first, the pace of growth accelerates with the least amount of warmth or longer days, so that flowers open before most other flowers are getting started. This gives them a significant strategic advantage, and as the flowers are full of nectar, they're terrific lures to early bees.

Whether California ever had comparable annuals in its grasslands is debatable, for few of our own native wildflowers are so prolific so early in the season. Perhaps in some locales, buttercups, milkmaids, and shooting stars were early-flowering substitutes. When nonnative flowers are able to usurp the range of native wildflowers we say that they have become naturalized, meaning that they behave as though they were an integral part of the original landscape.

The period over which mustard and radish weed continue to bloom depends on how warm the spring becomes; if all

# California Poppy

*Glenn Keator*

Many historians have noted that although the metal gold first attracted large numbers of people to California during the Gold Rush of the mid-1800s, the real gold lies in the land. This truth is revealed by the fertility of California's soils but is literally true for the mantle of golden poppies draping the hills and bluffs in spring. So vivid were early displays of poppy fields that mariners commented that they could see poppy gold from far off shore. Today the golden fires have been damped in most places, but occasionally we have a year in which for a little while we can relive the way the poppies used to be.

No finer or more representative wildflower could be chosen for the state flower than the California poppy. California poppy occurs not only along coastal bluffs but throughout most foothills both in the Coast Ranges from Oregon to Baja California and along the western front of the Sierra Nevada.

A short-lived perennial, California poppy springs from a long, carrot-colored taproot, with a close cluster of much-divided, feathery, bluish-green, fernlike leaves. The tulip- to saucer-shaped flowers are carried over a long period of time—one by one—above this foliage. Blossoms may appear as early as March and continue well into early summer, but with surprising changes in size and color. Earliest flowers are large, vivid orange,

is cool, with intermittent showers, the display may last for well over a month, but when temperatures suddenly soar as they sometimes temporarily do, the plants may quickly go over to fruiting and seeding.

By spring's end, both have made myriad seed pods that turn from green to brown and split open along two sides. Winds shake the brittle stems back and forth to aid the seeds in spilling all around the parent plants. Soon nothing is left, and so by summer, field mustard and radish weed are hardly to be noticed, for the old stalks have fallen down and died.

More than one kind of mustard is naturalized along the coast. Although field mustard is by far the most widely distributed and by far the showiest, another version called black mustard (*Brassica nigra*) opens its smaller yellow flowers about the time field mustard is going to seed.

Unlike its showy relative, black mustard flowers stintingly, with few flowers open at any one time, but grows taller and broader as the season advances (at least near the coast). By the time it has finished, the plant may stand four to five feet tall with an almost equal spread—much like a small bush—the many branches having borne a long succession of flowers. Flowers in mild summers may continue through to at least mid-summer before plants begin the slow process of seed ripening and browning that is inevitable with annuals.

Even by summer's end and into fall, the old skeletons of black mustard remain in place, unless some unusually strong wind should come along and topple them over, then carry the plants in a clumsy act of rolling. Some close relatives to black mustard—particularly the tower mustard (*Sisymbrium altissimum*)—make excellent "tumbleweeds," because of this ability.

POPPIES

GLOW ON DIM FOGGY DAYS...

and tulip-like; late flowers are smaller and shallower, and often a paler yellow or yellow-orange. In addition, coastal forms on beaches and sand dunes often differ by their shallower flowers—yellow centered orange offset by silvery foliage, colors which glow on dim, foggy days.

California poppies belong to a genus named for the German botanist Escholz, who presumed to create a better reputation for himself by the affectation of doubling the letters in his name to Eschscholz! Most eschscholzias are strictly Californian; there are several other kinds, but all except the California poppy are annuals with smaller, weaker roots, and all lack the pretty, red-tinged rim below the petals (the receptacle of the flower) except for California poppy. Other eschscholzias bloom quickly from mid- to late spring and then are gone; they range from the grassy rolling foothill country to the sands of the Mojave and Colorado deserts.

California poppy shows its allegiance to the poppy family by the pointed sepal cap which falls away as the flowers open, by the crinkled appearance of the petals, by the numerous stamens, and by the single pistil. Since poppy flowers don't offer nectar to insect visitors, they compensate by offering prolific pollen (hence the numerous stamens), an incentive to bees who use the pollen to feed their young. Distantly related to other native poppies, including the gaudy Matilija poppy of southern California and the delicate wind poppies of lowland woodlands, the group called the bleeding hearts are actually more closely related to those other poppies than is the California poppy. Bleeding hearts have similar, fern-like leaves, but their flowers—designed for bumblebees and large beetles—use a very different design approach: nodding, pink, heart-shaped flowers with stiffly clasping petals, the stamens and pistil hidden inside and the nectar deeply buried at the top of the heart.

California poppy was quickly adopted by the early English gardeners and brought into cultivation as a garden flower. Through the years, hybridizers have created a wide range of colors seldom seen in wild poppies—pinks, white, cream color, and rose—and forms with doubled petals. Although these introduce variety in the garden, the deep orange-gold flowers of the original wild form are still the most beautiful. Because most poppies contain opiate-type alkaloids, they have potential medicinal use. Perhaps our state symbol will one day be hailed for its curative properties as well.

*Linda Yamane*

California Poppy
    speaks
        strongly
   in rich
      orange
          tones
  singly
      or in concert
 on blazing
      hillside
 or meadow.
But did you
  know
      the magic

and
      medicine
found within?
Wintu and Yuki
  knew
      to put a bit
of fresh root
in the cavity
  of an aching
      tooth.
Ah, relief!
And Rumsien Ohlone
  parents
     knew
 that a flower
   or two
 underneath the
  bed
 would surely
   lull
 a young one
into quiet sleep.

Never underestimate
 the
 power
 of a flower!

# Yerba Buena

*Glenn Keator*

Our accepted common name for this creeping mint derives from the Spanish for "good herb," an allusion to its curative and fragrant principles. So abundant was yerba buena along our coast, particularly on fog-drenched headlands and knolls, that the original name for San Francisco was Yerba Buena, and we still have Yerba Buena Island in the middle of the Bay.

The plant seeks the solace of dappled shade, where scrublands meet forest, or where grasses thin along steep, stony embankments. It keeps company with selfheal, another mint relative; stonecrops; dudleyas; coastal succulents; woodland strawberry; and various small ferns, such as California polypody. Once established, yerba buena spreads rapidly by rooting as it grows; the entire plant seldom exceeds an inch in height! Native plant gardeners are beginning to use this fragrant mint relative as a modest groundcover, in shaded places where summer water is infrequent, as, for example, under stately oaks. It is removed from its native haunts with difficulty, but responds when satisfied of conditions in its new home.

To recognize yerba buena, look for purplish-tinted, squarish stems bearing pairs of oval, daintily scalloped, light green leaves infused with the odor of mint. The tiny white flowers bear the stamp of the mint family—petals in a two-lipped arrangement, with a "throat" between—and appear sporadically in spring between leaf and stem. Except for a brief winter rest, yerba buena may be found throughout the year.

Yerba buena belongs to a generic group of far-flung distribution, for the very same genus *Satureja* contains the culinary herb, savory. Unlike yerba buena, however, the savories are small, bushy plants, with numerous, weakly ascending twigs forming dense clumps

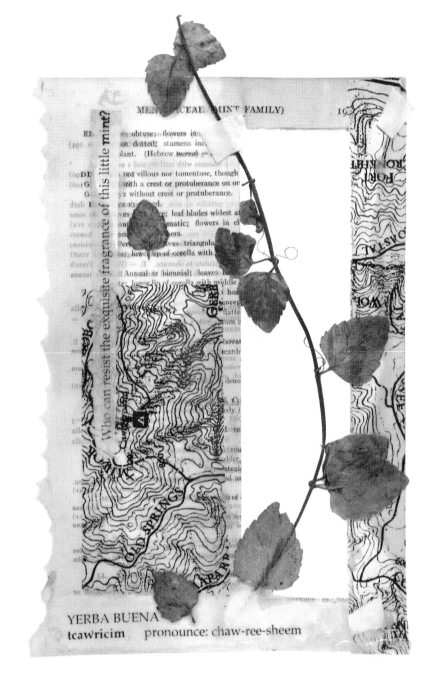

Who can resist the exquisite fragrance of this little mint?

YERBA BUENA
tcawricim          pronounce: chaw-ree-sheem

rather than crawling along the earth. It's only when you see the similarly-shaped white to pale purple flowers that the relationship becomes believable. Savories come to us from Mediterranean lands of southern Europe, where their use as an herb of healing harkens back to prehistoric times.

*Linda Yamane*

It is impossible for me to think of yerba buena without thinking also of my grandma, my father's mother. I could make her squeal with happiness by bringing her cilantro, piñon nuts, or yerba buena. If it was cilantro, she'd make "a'bondigas," a delicious soup filled with meatballs made with rice and plenty of cilantro. We'd sit together at her tiny table in her tiny kitchen, warmed from the inside out by her flavorful soup and lively conversation. If it was "piñones" I brought, we'd feast then and there, marveling at the magnificence of those smooth and creamy kernels. If yerba buena, then it was time for tea! We'd make it and drink it and sit around feeling good.

Who can resist the exquisite fragrance of this little mint? Even if it didn't taste so wonderful, the smell alone would be worth the effort of brewing tea. No wonder my grandma loved it so. In fact, they were a lot alike—both tiny and delicate in appearance but actually remarkably sturdy and capable of thriving on few amenities.

To this day I feel immensely happy when out walking to look down and find yerba buena growing. At the very least I reach down to rub a leaf between my fingers then stand for minutes inhaling its wondrous scent. If it is plentiful and gathering is permitted, I may snip off lengths to take home with me, remembering gathering sprees with other Rumsien friends. We stooped and squatted among the greenery, breaking off snippets here and there and reminiscing about our loved ones, mothers or grandmothers, who also gathered and prepared yerba buena tea. Later we'd wash it and wrap it in little bundles, ready to dry and just the right size for the next cup of tea.

# Cattail

*Linda Yamane*

Where cattails grow, you know there is water—and the raw material to satisfy an assortment of human needs. In California, the cattails and their frequent companion, the tules, have been used in a variety of ways.

In times past, when wetlands were more numerous than they are today, great volumes of slender cattails and tules were used as insulating thatch for structures, both large and small. Tule houses were common throughout many parts of California, though during the warm months people preferred living outdoors, using their houses primarily for storage. I've never lived in one, so I can't vouch for their comfort on a long-term basis, but I enjoy being inside them whenever I get the chance. I love the smell—it's like being inside a basket. Willow poles, arched and anchored into the ground and secured at the joints with strips of willow or other bark, create a dome-shaped framework. The walls are then thatched with mats of tule and secured to the willow frame. Walls can be made surprisingly thick and mats that overlap like shingles will easily shed water. A smallish door in one wall and opening in the roof allow folks in and smoke out. Of course, styles and construction details varied from place to place, varying from large to small, being circular or elongated, but in the old times villages of clustered tule houses dotted the landscape throughout much of California. Cattails or tules were also held together by twining to form mats of varying sizes—for sleeping, sitting, working, entertaining, covering doorways, for shade, and a myriad of other uses. They can be made quickly or with meticulous care and design, depending on their use.

Lengths of cattails were twisted and plied into rope or other size cordage, and cattail rope was used in some areas to bind bundles of tules into the tule boats characteristic of many Native California communities. Air pockets in the interior of the stems provide the buoyancy that makes tules an ideal boat-making material. I helped make a fifteen-foot tule boat once and recall the work involved. I remember laying out preliminary small bundles that were wrapped and then bound together to form the larger bundles needed to construct the boat. We lashed two large bundles together by figure-eighting them to a third small central bundle,

and this became the base of the boat. We then lashed another bundle atop each side to form the gunwales, taking special effort to lift each end into an upward sweep for ease in cutting through the water. The test-ride was an experience I will never forget—the solidness and confidence with which it rested and moved through the water, the feeling of connection to the past, and the mending of a cultural thread that had been so long broken.

Some of my favorite baskets are those made of twined tule or cattail by the Klamath and Modocs of northern California and southern Oregon. These baskets are wonderfully sensuous to hold because they are so flexible. Their flexibility comes from the fact that the warps, the radiating foundation elements, are of fine, two-ply tule or cattail cordage, rather than rigid warp sticks such as willow or hazel. Intricately patterned hats, bowls, gambling trays, quivers, burden baskets, and other forms were all skillfully fashioned from these prolific wetland plants.

Imagine the security of being able to make a house wherever and whenever one was needed, from materials readily available, and requiring technological skills available to any average adult in the community. The thought certainly stands in stark contrast to the threat of homelessness experienced by many modern people. Imagine, too, living in a home of the same texture and colors that surround you. Imagine working outside a tule house sitting on a tule or cattail mat, wearing a tule hat, surrounded by baskets of different shapes and sizes, all made of cattail or tule, and perhaps even eating a meal including the thick starchy cattail

rhizome. Few areas revolved so completely around these two companion plants, but wherever they grew, they were used and appreciated for that which they offer.

*Glenn Keator*

Just as willows announce the presence of permanent water courses, so cattails proclaim the presence of marshes and shallow freshwater lagoons. There they occupy a zone just beyond the water's edge, where roots and rhizomes are buried in permanently soggy mud and shoots emerge a few inches through standing water. In this zone they vie with another grasslike plant called tule (actually a member of the sedge family). These two plants together—cattail and tule—nearly exclude all others by growing tall and fast to shade out potential competitors. Even though a marsh would seem a likely place for plants to thrive, few plants are adapted to this environment for a very important reason: little oxygen dissolves in water or soggy soil. Oxygen is what keeps roots healthy, since they require oxygen to burn food for energy, or respire, just as leaf and stem cells do. In fact, it's little appreciated that plants need and use oxygen in a manner identical to animals; foods such as sugars and fats are broken down in the presence of oxygen into carbon dioxide and water in order to produce the energy which fuels cell activity and makes life possible.

Marsh plants adapt to the lack of oxygen by ingenious devices: cattails use the honeycombed air chambers inside their long, flattened leaves to carry oxy-

gen to roots and rhizomes in the mud. In tules, the air travels in tubes through the green stems, in reality the same basic principle. These air chambers render cattail leaves and tule stems useful for temporary thatching on simply constructed shelters whose framework may be willow branches coming from adjacent areas. Because these air chambers create dead air spaces, just like those designed into modern insulation, cattail leaves and tule stems keep enclosures cooler on hot days, warmer on cold nights.

Cattails are easily identified by their long, bluish-green, grass-like, flattened leaves; but at the end of fall, these die back and the foods manufactured during the long summer days are sent to shallow, underground rhizomes until warmer days return. These underground rhizomes thus become a repository for starch, much as it's stored in the tubers of potatoes, for starch is a good way to store food: it consists of long chains of sugars hooked end to end, which are easily broken down for sustenance and energy or converted into other substances as needed when growth resumes.

The brand new shoots of cattails emerge in spring, and their inner "hearts" are delicate and flavorful at that time, with the taste of cucumber and the tenderness of fresh asparagus. As they unfold their fans of new leaves, a young stem appears in the middle carrying a double-fisted spike which slowly rises above the leaves high into the air. By the end of spring, this spike looks like a long, double-jointed sausage. The upper sausage opens to shed so much yellow pollen that the lake surface is blotted yellow, and great wafts of it spray through the air on windy days assuring its movement to other cattail plants. So abundant and nutritious is this pollen, that several peoples learned to harvest it for food. Meanwhile the lower "sausage" spreads thousands of delicate spidery stigmas to receive pollen. By mid-summer, the upper spike disintegrates, its job done, and the lower spike slowly turns from green to brown. By fall, the brown female spike is ready to release the thousands of nearly microscopic, hair-covered seeds. Gusts of wind and rain squalls help the process of opening these spikes: the way the seeds and their white hairs are packed together is especially ingenious, for as the spikes open, they expand greatly as though exploding from internal pressures, and great masses of white-plumed seeds are scattered far and wide. Long-distance dispersal once again allows efficient transfer of seeds to widely separated marshes and ponds.

# Sedges

*Glenn Keator*

California has no fewer than 140 species of sedges, most of them inhabitants of our shrinking wetlands: in montane meadows, along streams and rivers, bordering marshes, bogs, and swamps. Sedges superficially resemble grasses, but are relatively easy to distinguish by the following traits: sedge leaves

have a central channel (furrow) running down the middle of the leaf its entire length, and a corresponding midrib (keel) on the backside of the leaf; sedge stems are triangular or three-sided and solid inside (although there are tiny air pores giving the tissue a porous appearance under a hand lens); and sedge flowers are unisexual, the male flowers borne in slender spikes above the female flowers.

Native sedges grow in one of two patterns: discrete clumps that gradually make new clumps close by (old plants consist of many densely clustered leaf rosettes); and clumps strung out over a large area and interconnected by creeping underground stems called rhizomes. The latter were important to Native Americans for their rhizomes (usually referred to in the literature as roots), for they contained long, strong fibers that were widely used to weave finely crafted baskets. Only those rhizomatous sedges that grew in sandy places were generally used because their rhizomes grew straight and true. Sedges occurring in rocky places—by contrast—had crooked, sinuous rhizomes poor or unusable for fibers because of their shape.

Sedges join another grasslike group in their preference for a permanently high water table: the rushes (*Juncus* spp.). Rushes too are interesting for their varied structure and specially designed flowers, but their long, stiff stems were not of high quality for basketry and fibers.

Rushes come in three basic forms: small annual plants with narrow leaves and minute flowers (look for these in small depressions that fill with water in

winter and spring called vernal pools); perennials with stiffly upright, cylindrical green stems terminating in umbellike clusters of flowers and a rounded, stiff leaf with sharp tip; and perennials with flattened, sword-shaped, iris-like leaves and separate stems bearing heads or umbels of flowers. The several species with stiff cylindrical stems and sharply pointed leaves are most often noticed, for their texture and pattern is easily distinguished from grasses or sedges. The iris-leafed kinds always come as a surprise, for out of flower they closely resemble leaves of the iris family. If you hold a leaf up to the sky, you'll see irregularly spaced dark lines (inner cross hatchings) running across the leaves on the inside; you'll also notice a white membrane along the lower third or so of the inner leaf edges. Actually, each leaf is a broad leaf folded in half, so the inner edges are still separate while the outer edge is at the point of the fold.

Rush flowers represent a partway stage in the evolution between insect-pollinated flowers and wind pollination. Each flower (use a good hand lens) consists of six bronze or whitish petals surrounding three or six pale yellow stamens and a single central pistil with three long, feathery, raspberry-pink stigmas. The stigma design together with the abundant dry pollen suggests the main mode of pollination here is already by wind, but the retention of unnecessary petals is a leftover from times when the ancestral plants had colorful petals to attract insect visitors.

Both sedges and rushes are general indicators of high water tables, for few

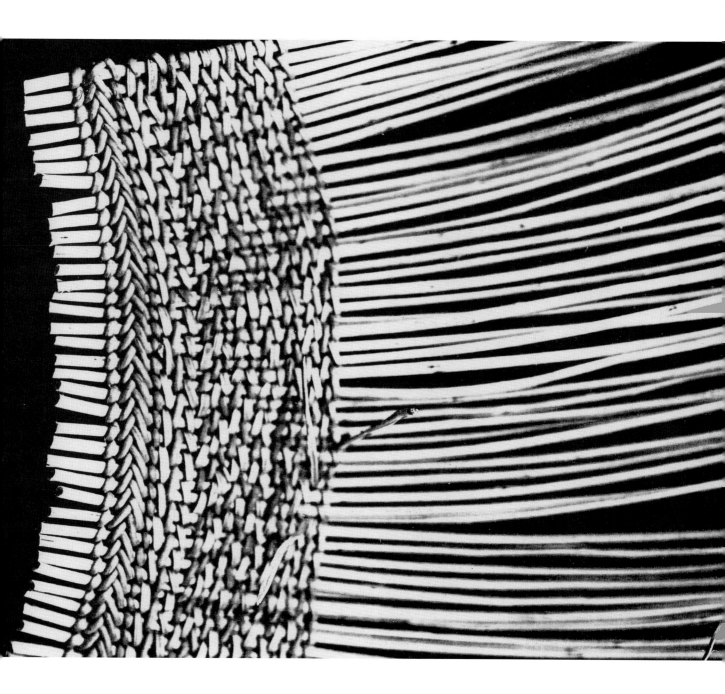

*"When you gather, you always pray for the plant and the land. And when you're praying for the plant and the land, you kind of make a deal with it, saying that it's going to live on, and that one day it will be a beautiful basket. Then, when you take it home you still have that agreement, so that helps you to clean it and do it right, instead of just taking it home and getting lazy."*

—Josephine Conrad, Karuk/Wiyot basketweaver

❦

*Mabel McKay with sedge roots.*

kinds will grow where roots have to endure drought or penetrate deeply to tap water. Because they live in wet to soggy soils, they have developed adaptations to the low oxygen content in these soils by delivering air to the roots. Sedge stems are spongy with honeycombs of air channels that extend clear to the underground rhizomes and roots; rush stems are pithy with similar air chambers, or their leaves are hollow in the center to allow free passage from tip to base, where the air is passed on to roots.

Because sedges and rushes aerate their roots they can live where there's little competition with other plants. Their fast growth, rapid vegetative reproduction, abundant seed set, and efficient air circulation guarantee ecological niches few other plants have managed to adapt to so well.

*Linda Yamane*

Admired throughout the world for their exquisite workmanship and beauty, California Indian baskets are a source of pride and cultural continuity for Native California people today. Central California baskets, such as those made by the renowned Pomoan weavers, owe their existence in great part to a little-noticed plant—sedge. In coiled baskets, it is sedge that wraps around the foundation rods and, except for the darker pattern materials, it is sedge that we see.

Sedge has been used by generations of California Indian weavers, and is undoubtedly one of the most valued and widely used of the basketry plants. It was only after I began traditional basketry that it became a part of my life. I'd

watch for clusters of slender green leaves, rising up, then emanating outward and relaxing into gentle arches. From a distance, I was sometimes fooled by clumps of tall grasses, but up close I'd look for sedge's unmistakable distinguishing feature—a distinct wedge running the full length of the center of each leaf. "Sedges have wedges" was a rhyme I had learned years earlier and it now became relevant as I sought to confirm each find. It was important to begin locating this vital plant—for without it the baskets could not be. This problem of diminishing resources and restricted access to plants has become a serious threat to the continuance of basketry and other Native traditions.

Sedge has variously been called "cutgrass" or "sawgrass" and if you've ever handled the foliage you'll understand why. The leaves can easily slice the skin, and experienced weavers, when digging sedge roots, wear long sleeves to avoid the nicks that are inevitable when working bare-armed.

Offspring plants eventually sprout from underground runners, or rhizomes, produced by mature plants, and thus sedges establish themselves in beds of varying sizes, often intermixed with blackberry or poison oak. Weavers dig for the rhizomes which are easily distinguishable from roots of companion plants by their fibrous outer layer. Once located, the weaver then follows the runner in both directions, clipping it off at the base of both the parent and offspring plants. This leaves the two plants healthy and intact, each capable of sending out new runners the next season. Weavers need to be able to work the same sedge

beds repeatedly, for it takes years to establish the conditions that generate the finest roots. Beds that have been regularly cultivated yield long, straight rhizomes because harvesting keeps the area thinned and uncrowded. In uncultivated beds growth is restricted and the soil more compact, resulting in short, gnarled rhizomes little suited for weaving.

Soil conditions also determine qualities such as color, strength, and pliability. Sedge growing in sandy conditions develops rhizomes that are the whitest in color. Loamier soils will result in creamier colors, and rhizomes growing in heavier dirt or leaf litter will be brown.

After harvesting, the rhizomes are split lengthwise, stripped of their bark and seasoned for several months before being split again and trimmed for weaving. The meticulous processes of harvesting and preparing materials can involve as much or more time than the weaving itself, and bring the weaver in close relationship with the plant world. Older weavers were careful how they talked to the roots, being sure to compliment them even if they were a bit short. Sometimes it was necessary to store certain roots separately because they didn't like each other. Roots that are spoken well of are more likely to come up long—and some weavers thought it unwise to brag lest they come home empty-handed. There have always been rules for gathering and handling of plant materials, and the rules for basketry plants have varied from area to area. For many, there are sexual or menstrual restrictions, the singing of songs, and the giving of gifts to the ground. Sedge digging can be a

lone venture or group affair, but whether solitary or social it always ends with some form of "thank you" for that which is given and gratefully received.

Baby baskets, food bowls, cooking baskets, storage baskets, gift baskets, gambling trays, winnowing and gathering baskets, ceremonial baskets, funerary baskets—from birth to death, baskets have been an intrinsic part of traditional Indian life. We need the plants to make the baskets, and when we use them we honor them. We take care of them and thank them—and give back to them. When we honor the plants we honor our ancestors. When we make the baskets, we keep our connection to the past alive for the future.

# Grasses

*Glenn Keator*

Grasses are among our most successful flowering plants even though few people think of grasses as flowering. The thousands of species have evolved to inhabit unique niches on the planet. Grasses may be responsible for the evolution of grazing animals and they may even be implicated in human evolution.

How do you recognize a grass? Here are things to look for:

Grasses have leaves divided into long, usually narrow blades; each blade has a sheath where it joins and encloses the stem, and a tonguelike ligule at the junction of blade and sheath.

Grasses have rounded stems that are hollow in the middle and sometimes jointed as in bamboos.

Grasses bear spikes, panicles, or racemes of tiny flower clusters called spikelets. Each spikelet consists of a pair of green to brown floral bracts (glumes) above which one to several greenish, brownish, or reddish florets (tiny flowers) are produced.

Grass florets consist each of another pair of ensheathing floral bracts (the lemma and palea), three stamens, and two long feathery stigmas attached to a one-seeded ovary.

Grass seeds are fused to their ovary to form fruits we call grains. Grains are often aided in their dispersal by attached barbs or spines called awns, or their large, food-rich grains are harvested by animals and cached for later use.

Other groups of grasslike plants include the sedges (family Cyperaceae) and rushes (genus *Juncus* in the family Juncaceae). All three groups are designed for wind pollination, hence their lack of colorful flowers.

Why, when insect and bird pollination seem so successful in streamlining reproduction, have these flowering plants returned to wind? Wind pollination seems to develop in habitats that are especially subjected to windy conditions, winds constant enough to be relied on for successful pollination. Stream courses are typically windy, so it is logical that many trees along such corridors are wind pollinated. Open fields are another habitat where winds may blow unimpeded over large areas. Finally, the time of the year may have something to do with it:

winds are more reliable during winter and spring storms. Grasses seem to have responded to such windy conditions by living mostly in wide open spaces, and often flower in winter or spring when winds are commonplace.

What changes have occurred to bring about wind pollination? Grasses have lost all traces of normal sepals and petals, not wasting energy by producing colorful and fragrant flowers. Another energy savings is the elimination of nectar, the reward offered to insect visitors. Stamens have grown extra-long stalks to position their anthers beyond the rest of the flower so that when pollen is ripe and ready for its journey, it is directly shed to the wind. Pollen grains are smooth and dry so that they do not stick together (the opposite of insect-pollinated pollen). Each pollen grain can make its own journey separately and independently from the others. It's this quality of wind-borne pollen that makes it so effective at bringing on those allergies called hay fever.

Stigmas also extend beyond the flower when their turn comes. Stigmas are like feathers, lined with myriad finely branched hairs that are effective at trapping and catching pollen from the air. They are hung out at a different time from the stamens in order to prevent self-pollination.

Other interesting features of grasses include their four modes of overall growth as well as the way in which leaves regenerate. First the growth patterns:

*Annuals.* Most of our present grasslands are filled with annual grasses. These come up soon after winter rains

start, completing their flowering and seeding by the time soils dry in late spring. California's classic cycle of green and brown is the result of millions of annual grasses such as wild oats (*Avena* spp.), foxtails (*Hordeum* spp.), nonnative ryegrasses (*Lolium* spp.), nonnative bromes (*Bromus* spp.), annual blue grass (*Poa annua*), and nonnative fescues (*Festuca* spp.).

*Perennial bunchgrasses.* Most of

California's original native grasses—particularly in coastal areas—are perennial. Such grasses live many years, have deeply probing roots, and grow as ever-enlarging clumps. Although they take a summer rest, they never completely die back or turn brown. Our original grasslands must have had a very different aspect, particularly in summer.

*Rhizomatous grasses.* Several native grasses are rhizomatous; that is, they have shallow underground stems that

run and branch in all directions, sending up leafy shoots periodically as they spread. Some rhizomatous grasses make dense stands; others more open colonies. One particularly obnoxious coastal nonnative rhizomatous grass is dune grass (*Ammophila arenaria*), deliberately introduced to our beaches to stabilize the blowing sands. Dune grass grows so densely it excludes all other vegetation, while some of our own native rhizomatous grasses, such as dune ryegrass (*Elymus mollis*) make open colonies that allow spaces for many other species.

*Sodgrasses.* Sodgrasses are most familiar to city dwellers, for they are responsible for lawns. Sodgrasses grow as very dense, interconnected clumps that spread laterally by short underground stalks. Since these underground stalks immediately send up new leafy shoots, the resulting sod appears continuous, without gaps or holes. Few native grasses grow in this fashion, and virtually none of the grasslands around the Bay Area have sod formers.

How do grass leaves regenerate? Instead of showing finite growth once they've matured, grass leaves have special zones of growth (meristems, where new cells are constantly being created to add new growth) at their base. Were it not for this feature, mowing grasses could be done one time only, but because grass leaves keep adding more growth from their bases, they grow long again and constantly need to be clipped back. This attribute makes lawns laborious to keep up, but it is also the reason grasses have adapted so well to the animals we call grazers. Imagine grasses

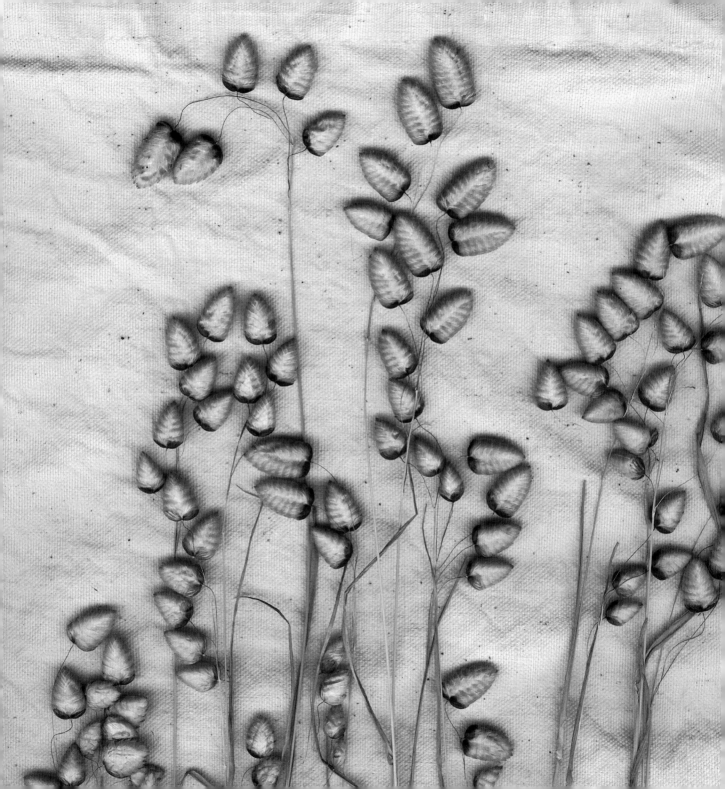

that couldn't regenerate themselves after grazing! It is due to this ability and none other that grazing animals were able to evolve. Of course, a trick such animals had to develop was a means for digesting grass leaves, since they're high in cellulose. Were it not for special cellulose-digesting bacteria that live inside the gut of such grazers, grasses would be impossible as food.

Grasses have also affected our own evolution. First, grazing animals became attractive sources of food for intelligent, widely roaming, upright apes. Second, grasses develop grains high in stored nutrients, including vegetable proteins, fats, and carbohydrates. Such grains are easily stored for long periods and complement flesh in the diet. Grass grains also lend themselves to cultivation on a large scale, making agriculture feasible. Grasslands also contain many other important seasonal food plants, since their habitat tends to exclude woody plants such as trees. Trees often shade out other potential food plants, since most food plants require high light energy to complete their life cycles. Such food plants include many bulbous and tuberous plants as well as a large range of plants with edible seeds. Most of our modern crop plants require open fields, not shaded places, for successful cultivation.

Let's turn our attention now to the role of grasslands in California's flora. No one knows for sure the extent of our original grasslands, for they have been managed by humans since time immemorial. Although the great Central Valley probably was home to vast tracts of open oak woodlands and grasslands,

other areas of extensive grasses are less certain. We don't really know if the rolling grassy hills we now see throughout our foothills or the grassy pastures along coastal bluffs were always there.

Native Americans probably first imitated nature by using fire to manage grasslands; there have always been lightning-caused fires, where bone-dry grasses caught fire. Soon Native Americans opened new areas by burning, for fire favors the regeneration of grasses over scrub (shrubs) or forest (trees). Since grass roots penetrate deeply, a surface fire merely stimulates them to make luxuriant new growth the following spring. The ash from fire actually fertilizes the grasses for improved vigor. Young shrub seedlings and tree saplings are killed outright by these same fires.

Why did Indians open areas formerly covered with trees or shrubs? There are several advantages: more open habitat for browsing animals (thus easier to hunt as well), more habitat for food plants, and more habitat for the grasses themselves, several of which served as food or basketry materials.

In the absence of fire, many of our present-day grasslands may revert to shrubs. Look, for example, at some of the areas where fire has been banished and grazing has been eliminated. Usually there are shrubs such as coyote bush (*Baccharis pilularis*) or yellow bush lupine (*Lupinus arboreus*) in evidence. If left this way for long, it's easy to see that the grassy cover gradually gives way to dense shrubs that choke off light to grasses and their companion plants.

Other changes to our grasslands are somewhat better documented—what has become of our perennial bunchgrasses, for example, is that in some areas, they're alive and well—often where grazing has been partially eliminated or where they were never crowded out in the first place, as on serpentine soils (where nonnative grasses cannot grow) or under oaks (where nonnative grasses cannot compete).

But for those wide-open areas on normal soils, most native bunchgrasses have been in decline until recently. The story goes something like this: when sheep and cattle were first brought in by the Spaniards, they preferred the taste of native bunchgrasses. Seeds of other grasses inadvertently accompanied the newcomers in bricks, soil on boots and other clothing, soil on implements, or mixed with seeds of crop plants (such as wild oats mixed with cultivated oat seeds). As more and more grazing pressure reduced the vigor of native grasses, the interlopers grew and prospered. Many of these nonnatives were preadapted to "disturbed" sites, so they proliferated quickly. Finally, under intense grazing pressure and steadily increasing competition from the newcomers, the native bunchgrasses died out to be replaced by the "aliens." Today we see vast tracts of nothing but nonnative annual grasses that renew the cycle of green and brown every year with the winter rains and summer drought.

Of the plethora of grasses you see along the coast, two draw special attention: pampas grass (*Cortaderia jubatum*) and rattlesnake grass (*Briza maxima*). Both are foreigners, pampas grass from Argentina and rattlesnake grass from the Mediterranean.

Pampas grass has been sold for decades as an ornamental grass, meaning that it creates special effects in gardens. You'll notice the immense flowing mounds of leaves year-round, but in flower, the huge, fluffy, white to pink plumes are unmistakable. Like immense feather boas, they create drama by their texture, color, and size. And they dry beautifully as nearly immortal bouquets. Yet such a seemingly impressive grass has become an agent of destruction as it "escapes" from gardens to menace native vegetation all along California's coast. Well adapted to drought and rocky promontories, its roots seek out all possible pockets of soil and push out any other plants in the process. These roots are strong and fibrous, like tough ropes, almost impossible to remove. And then those plumes produce even more seeds to start the process over again. What was once a prized ornamental has thus turned into a despised truant.

Happily, rattlesnake grass is less aggressive. Well adapted to coastal grasslands, this annual dies back with the rest of the imports, but seldom spreads so flagrantly that it supplants native wildflowers. Instead it provides a bit of drama as the spring season comes to a close, for then the nodding spikes of flowers begin to show themselves. Green at first, these gradually pass into tan, but if they're to be used in dried flower arrangements, pick them while still green; otherwise the spikes shatter. So much do these "spikelets" of flowers resemble the tails of rattlesnakes that further explanation for the name is unnecessary.

*Glenn Keator* is a teacher, writer, and botanist whose interest in plants was inspired by his grandmother's garden and family camping trips to the desert and the Sierra. He lives in Sebastopol, California. In addition to writing several field guides, Glenn is the author of two gardening books, *Complete Garden Guide to Native Perennials of California* and *Complete Garden Guide to Native Shrubs of California.*

〜

*Ann Lewis* is an artist and writer who makes one-of-a-kind books. Her work is held in the Getty Center for Arts and Humanities, Stanford University, and numerous private collections. Ann describes her approach to plants as "playful and unscientific" and is planning a forthcoming collection of visual and written reflections on plants of the South. She is currently living in Strafford, Vermont.

〜

*Linda Yamane* traces her ancestry to the Rumsien, the native people of the Monterey area. She has been active in retrieving Rumsien Ohlone language, songs, basketry, and folklore—cultural traditions that were once thought lost. She is an artist, writer, researcher, and cultural demonstrator living in Seaside, California. Linda is a contributing editor to *News from Native California* and was on the founding board of the California Indian Basketweavers Association.

**Artwork by Ann Lewis**

Title Page and Back Cover - collage
Copyright page - *Grassy*, collage with slide mount
Facing copyright page (detail) and page 70 - *And the Fruits Ripen, Cow Parsnip*, collage
Table of contents page - *How 'Bout Those Roots*, fern print
Page 13 - *Do Not Touch*, On Gathering, collage
Page 21 - *So Hairy*, photocopy of Lichen
Pages 23 and 25 - Seaweed prints
Page 26 - *The Sea is Deep*, seaweed collage with slide mount
Paage 29 - *Like a Flipper*, seaweed print
Page 31 - Fern print
Page 32 - *Bracken  and Woodwardia*, collage
Page 33 - *A Fern*, collage with slide mount
Pages 34 and 35 - Horsetail Fern prints
Page 39 - *From a Dead Eucalyptus Tree, Officer*, photocopy
Page 41 - *The Story of Blue Gum*, Eucalyptus, collage
Page 43 - *Witcher Willow Withes*, collage
Page 47 - Coyote Bush print
Pages  52,53 - *Why Take a Chance?*, Poison Oak, collage
Page 56 - *Five Leaves*, Blackberry print
Page 58 - *Stung by Taawax*, Stinging Nettle collage
Page 67 - *Douglas Iris: Of a Farflung Family*, collage
Page 73 - California Poppy print
Page 74 - *Glow on Dim Foggy Days*, California Poppies, collage
Page 76 - *Who Can Resist?*, Yerba Buena collage
Page 85 - *Pompous Grass*, Pampas Grass print
Page 86- *Is this Beauty a Grass?*, collage with slide mount
Page 88 - *The Grass Twins (or are they Weeds?)*, print

**Photo and Image Credits**

Front cover, facing table of contents page and page 3, Brenda Tharp. Inside front and back covers, Glenn Keator. Page 2, Dixi Carrillo. Facing copyright page and pages 13, 32, 41, 43, 52-53, 58, 66, 70, 74, 76, Ann's collages photographed by Sibila Savage. Pages 22, 46, 84, and 92 Richard Frear, National Park Service. Page 14, photographer unknown, 1925, Courtesy of Smithsonian Institution, National Anthropological Archives. Page 17, Spore Print by Tamia Marg. Page 27, Beverly R. Ortiz. Pages 44 and 82, Scott M. Patterson. Page 68, courtesy of Milwaukee Public Museum. Page 36, Jorge Vertiz. Title page, and pages 6, 50, Mark Klett. Pages 4, 5, 10, 64, and 78, Headlands Center for the Arts Archives. Page 60 - Soaproot Brush, made by Linda Yamane. Pages 40 and 81, photocopies. Pages 28, 42, 45, 61, and 87, scans made directly from plants.

## Plant List - Language Names

**Sp**=Spanish
**RO**=Rumsien Ohlone
**CM**=Coast Miwok

❦

### Algae/Seaweed

### Blackberry
*Rubus vitifolius*
Mora, Zarsamora (Sp)
'Een (RO)
Wate (CM)

### Blue Bush Lupine
*Lupinus arboreus*
Patito (Sp)

### Blue Dicks/Brodiaea
*Dichelostemma capitatum/Brodiaea*
species
Cacomite (Sp)
Rawson (RO)
Patchu, Waila (CM)
(wild onion—*Brodiaea* spp.)

### Blue Gum
*Eucalyptus globulus*

### Bracken Fern
*Pteridium aquilinum*
Manita (Sp)
Witt (RO)

### Bush Monkeyflower
*Mimulus aurantiacus*
Chupadero (Sp)

### California Poppy
*Eschscholzia californica*
Amapolla (Sp)
Cululuk (RO)
Munkai (CM)

### California or Woodland Strawberry
*Fragaria vesca californica*
Maduce (Sp)

Aium (CM)

### Cattail
*Typha latifolia*
Tule Ancho (Sp)
Xaal (RO)

### Chain Fern
*Woodwardia fimbriata*

### Common Horsetail
*Equisetum arvense*
Cañutillo (Sp)
Caax (RO)
Suk-ki'-yuk (CM)

### Cow Parsnip
*Heracleum lanatum*
Rama del Coche (Sp)

### Coyote Bush/Brush
*Baccharis pilularis*
Chamiso (Sp)
Puusen (RO)
Tcu-u (CM)

### Douglas Iris
*Iris douglasiana*
Cebollin (Sp)
'Uuner (RO)
Lawik (CM)

### Manroot
*Marah fabaceus*

### Monterey Cypress
*Cupressus macrocarpa*
Cipres (Sp)

### Monterey Pine
*Pinus radiata*

### Mushrooms

### Mustards
*Brassica* species
Wild Radish (Radish Weed)
*Raphanus sativus*
Rabano (Sp)

### Poison Hemlock
*Conium maculatum*

### Poison Oak
*Toxicodendron diversilobum*
Yedra (Sp)
Nissis (RO)

### Scouring Rush
*Equisetum hyemale*
Cañutillo del rio (Sp)

### Sedges
*Carex* species
Zacate (Sp)
Xuyxuy (RO)
Kici (CM) (pronounced "kee'-shee")

### Soap Plant/Soaproot
*Chlorogalum pomeridianum*
Amole (Sp)
Torrow (RO)
Haka (CM)

### Stinging Nettle
*Urtica dioica*
Ortiga (Sp)
Taawax (MO)
Sek (CM)

### Willows
*Salix* species
*Salix lasiolepis* (Arroyo Willow)
Saux de la Oja Finita (Sp)
Tarxasan (RO)Tewut (CM—
Bodega)/
Tewoot (CM—Tomales)
*Salix hindsiana* (Sandbar or Gray
Willow)
Sauz Cenizo (Sp)
Pays (RO)

### Yerba Buena
*Satureja douglasii*
Yerba Buena (Sp)
Tcawrishim (RO)